THE PLEASURE SHOCK

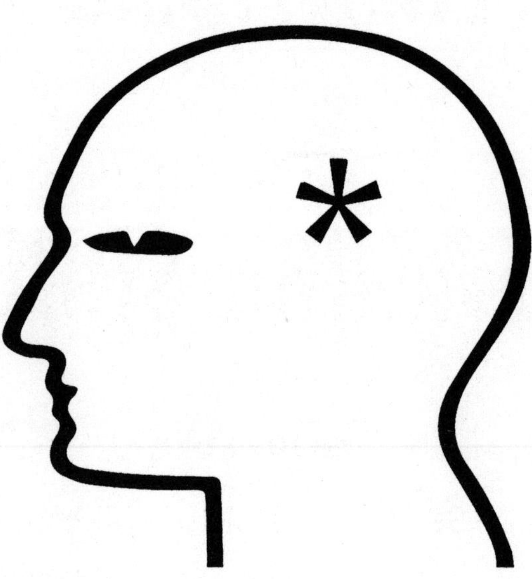

LONE FRANK

The PLEASURE SHOCK

The Rise of DEEP BRAIN STIMULATION and Its Forgotten Inventor

DUTTON

DUTTON

An imprint of Penguin Random House LLC
375 Hudson Street
New York, New York 10014

LIBRARY OF CONGRESS CATALOGING-IN-PUBLICATION DATA

Names: Frank, Lone, 1966– author.
Title: The pleasure shock : the rise of deep brain stimulation and its forgotten inventor / Lone Frank.
Description: New York, New York : Dutton, [2018] | Includes bibliographical references.
Identifiers: LCCN 2017029957| ISBN 9781101986530 (hardcover) | ISBN 9781101986547 (ebook)
Subjects: | MESH: Heath, Robert G. (Robert Galbraith), 1915–1999. | Deep Brain Stimulation—history | Mental Disorders—therapy | Physicians—history | History, 20th Century | Louisiana
Classification: LCC RC350.B72 | NLM WM 11 AL6 | DDC 616.8/046—dc23
LC record available at https://lccn.loc.gov/2017029957

PRINTED IN THE UNITED STATES OF AMERICA
1 3 5 7 9 10 8 6 4 2

BOOK DESIGN BY AMY HILL

For Morten Malling
My love—always

If it was possible to become free of negative emotions by a riskless implementation of an electrode—without impairing intelligence and the critical mind—I would be the first patient.

His Holiness the Dalai Lama
Society for Neuroscience Congress, November 2005

CONTENTS

THE PLEASURE SHOCK

Prologue

The doctors tell him to forget all about the wires. There are four of them, dragging from the back of his head to the floor and under a thick curtain into the next room. Ignore them, the doctors say, relax—concentrate on the task at hand. That won't be easy; he's never been with a woman before. And though they've been aiming for this moment, this *encounter*, for months, he can't shake the thought. *This feels utterly wrong.*

He glances around. The doctors have done the best they could with the sterile white room. They've put up some curtains and laid out a few rugs to give it something of a homey ambience. They've also softened the harsh overhead lights, which helps put him at a slightly greater ease. Of course, they've placed a bed in the room, even if it's just a regular white metal-frame hospital model. He can't avoid seeing her, sitting there on the bedcovers. She pats them politely and invites him to sit down beside her. He knows it is more like a summons, a requirement of his involvement in this experiment. He collects his wires and drags himself over to the bedside.

They have two hours. That's how much time she's been paid for, he's been told. She tells him to relax, take it easy. There's no rush. Why doesn't he start by telling her a little about himself?

He considers her more closely now. She has already taken off her dress, but fortunately she still has her undergarments on. She tells him that she's just turned twenty-one, but he senses that she is somehow older than her years. He's twenty-four, but right now he feels much younger, less experienced. He has a hard time looking her in the eye. When he finally manages to speak, he addresses the room, softly at first, then with more force. It's as if everything, his whole life, comes pouring out of him at once, the problems with his family, the drugs, his covert meetings with older men, the disgust he feels for what he is . . . and for what he is not.

As he talks, the woman moves closer and, though he does not notice when or how, she slips off her bra. It's lying on the floor like a deflated balloon. It's uncomfortable to look at her naked body but also strangely exhilarating. Finally, he senses the thing that brought him here. That feeling the scientists have managed to give him over the months he has been coming to the lab. A wave of warmth—*is it desire?*—spreads downward and tugs at him.

He thinks the experiment may be working, and this gives him the courage to stretch a hand out to her. He fixes his eyes on the pattern of one of the curtains, lets his fingers come into contact with her unfamiliar body. Her skin is surprisingly warm, softer and more yielding than what he is used to. He registers a few syllables of her voice and lets his hands roam.

Now it becomes a kind of dance, one that dances itself. But every so often, he is struck by a thought, almost as if someone is prodding him on the shoulder, telling him, urging him, to keep going. Intermittently, he remembers the wires and the room next door. There are four people in that room, and they are observing everything he does, recording every movement he makes, though they can neither see nor hear him, at least not directly. Right now, they are focused intently on the readings from the nine electrodes placed deep within his brain. With each movement of his hand, each sensation of skin on skin, signals from a densely packed area of gray matter run through the cables into the doctors' sensitive devices.

Initially, as I read the account of the experiment with B-19, I thought it was a sick joke, a parody of mad scientists run amok. But the more I read, it became clear: A psychiatrist had actually tried to convert a homosexual man to heterosexual preferences through brain stimulation, and both the doctor and his patient genuinely wanted the experiment to succeed. Electrodes had been implanted in the man's pleasure center and were stimulated while the man underwent a monitored "session" with a female prostitute. As unlikely—not to mention unethical—as the experiment seemed, the original article filled seven pages of an academic journal, the *Journal of Behavioral Therapy and Experimental Psychiatry*, in 1972. It was all there, in black and white, though the text and graphs left a lot to the imagination.

The language of the paper was sober and academic, which

made it all the more jarring. "Pleasure has long been known to be a primary reinforcing condition for acquiring and establishing behavior in animals and man," the preamble stated matter-of-factly. "Over the past several years considerable interest has also fastened on the fact that a pleasurable response can be induced by direct activation of the brain and raised hopes that this might be applied to the treatment of disordered human behavior." Were they really exploring a treatment for sexual orientation a few years after the Stonewall riots kicked off the gay liberation movement? Yet the scientific logic was rigorous and stringent. The treatment regimen was subtle and sophisticated. The encounter between the "patient" and his "treatment"—the prostitute—was designed as if it were normal lab practice.

In the same terse medical language, the paper provided a dry diagnostic sketch of the experimental subject. "Patient B-19 is a 24-year-old single, white male of unremarkable gestation and birth whose immediate family consists of parents, age 55, and a sister, 19," it began. But this laconic introduction belied the wretched circumstances of the patient's life. B-19's mother was a cold fish—in fact, B-19 could not remember receiving a single hug as he grew up—but his father was even worse. The man was an alcoholic, a violent, quick-tempered tyrant who was chronically disappointed in his depraved son. B-19 could muster positive feelings only for his sister, the only person with whom he felt comfortable sharing worries and disappointments.

B-19 seemed normally gifted, perhaps even above average, according to the psychiatrists. But he had suffered several suicide attempts and hospitalizations since his teen years, and was fixated on his past mistakes and deficiencies, so much so that he struggled to put his psychological life in order. He was a hypochondriac and horribly afraid of pain and dying. He could not tolerate criticism of any kind, could barely stand being with other people, and felt isolated from the world around him. Moreover, B-19 was apathetic, suffered from chronic boredom, lacked motivation, and generally had a deep-seated sense that he was wrong, valueless, and useless. Paradoxically, B-19 was also convinced he was quite special, a person God would reward for having endured so many cruel difficulties. "The extent of his paranoia fluctuates situationally, but is often of true psychotic proportion," the paper concluded.

Finally, there was his sexuality. That was what B-19 blamed for his desire to take his own life. He had dropped out of high school and joined the military, hoping to make a fresh start in life, but was discharged after only a month for his "homosexual tendencies." For a few years afterward, he roamed the country, hooking up with a variety of men. By the time he walked into the lab to have the cables attached to his head, B-19 was telling the psychiatrists that he was disgusted by himself and was no longer capable of experiencing happiness or pleasure. From anything.

I was appalled. But also curious. He reported feeling desire with the prostitute. How did this young man come to agree to have this untried, unorthodox, and frankly weird brain surgery? What did he feel as it was happening? What happened to B-19 after he left that room? And what exactly had the doctors done to his brain?

• • •

B-19 would be in his late sixties today. I knew, as a trained neuroscientist familiar with the conventions of studies with humans, that I would probably never learn how his life turned out. But there was another man at the center of this experiment, the scientist who conceived the treatment, built the devices, implanted them in B-19's brain, then published his findings. What was *he* thinking? He was obviously aware of his patient's pain, and seemed to want to help the young man by treating his unwanted urges. Gay Pride marches were being staged around the country, but the handbook of psychiatry—the *Diagnostic and Statistical Manual of Mental Disorders* (*DSM*)—still listed homosexuality as a pathology. The doctor and his patient almost certainly believed the patient had a mental disorder. Still, this was not the work of just any doctor trying to heal a patient.

The doctor's name was Robert Galbraith Heath, B-19 was not his only patient, and homosexuality was not the only disease he was aiming to cure. The references listed in his 1972 paper noted that over the past two decades, starting in 1950, Heath had implanted deep brain electrodes, which he also called brain pacemakers, in dozens of psychiatric patients with conditions ranging from schizophrenia to depression. I had to take a second look to believe it: Heath had started his experiments stimulating the brain several years before the first psychotropic drugs for treating mood and behavior came on the market. In 1950, the most popular treatments for psychotic patients were electroshock therapy and, in severe cases, surgical lobotomy. These were both crude interventions used with scant scientific rationale. In practice, they were performed idiosyncratically, with the procedures and the results varying greatly from doctor to doctor. Notoriously, some lobotomists simply hammered a sharp-edged instrument through the bone behind the eyeball and wiggled it around in the patient's frontal lobes, destroying neural tissue at random.

People were desperate for some way to help mentally ill patients. Fourteen years earlier, the Portuguese neurosurgeon António Egas Moniz invented the frontal lobotomy to treat schizophrenia. The procedure was considered such a resounding success that Moniz received a Nobel Prize for it in 1949. And within two years of the introduction of electroshock therapy in 1940, 4 out of 10 American hospitals administered the treatment. During the 1930s, there were 18,000 more mental health patients than available hospital beds in the United States. Care in asylums was mainly custodial, with a discharge rate of less than 15 percent. Being admitted to an asylum was a life sentence, and treatments were ineffective. Some patients received hydrotherapy, where the patient was forced into either very hot or very cold water, sometimes for days; or fever therapy, where patients were infected with the malaria parasite. These were supposed to calm down agitated patients. Others started to get insulin-coma therapy, which reportedly left the patients—if they survived—calm and responsive for a while.

The other option for treatment was psychoanalysis. Freud's view, which dominated American psychiatry in the first half of the twentieth century, dictated that mental health problems were not biological but rather the result of developmental disturbances triggered by events in early childhood that had been repressed and forgotten. For example, schizophrenia was a super-neurosis brought about by cold parenting, especially on the part of the mother. It's easy to imagine how most analytical psychiatrists at the time would have had a field day explaining patient B-19's various troubles by invoking the refrigerator mother and all-around unhealthy family dynamics, which Heath described in his 1972 paper.

That Heath, in this era and psychiatric context, had decided to treat mental illness by stimulating discrete parts of the brain made him a veritable pioneer. He had seen what lobotomy and electroshock therapy achieved—and found the aftereffects crippling. He was especially bothered by how lobotomy left most patients with a flat emotional life and deadened their social interactions. They might have calmed down, but their personality was stifled. At the same time, he did not accept that the brain was not part of the body. Like the rest of the body, its disorders and abnormalities could be treated with precision, if one knew how the brain worked. His clinic was based on the main tenets held when treating mental illnesses today: that biochemical imbalances and cellular disturbances cause them. He even hinted that heritable factors played a role.

Genetic studies have since revealed a substantial heritability to conditions such as depression, attention deficit hyperactivity disorder (ADHD), autism, bipolar disease, and schizophrenia. Schizophrenia is regarded as mainly a disorder of dopamine imbalance and treated with dopaminergic drugs, while depression is typically attacked with drugs that raise the level of serotonin.

(Serotonergic antidepressants were famously popularized in Peter Kramer's bestseller *Listening to Prozac* and have, since their introduction in the 1980s, been widely used to treat mood disorders as well as anxiety.) Heath was already thinking in this direction, but because neurotransmitters such as dopamine and serotonin had not yet been discovered, his approach had been to modulate regional brain activity electrically. In this way too, his research could be seen as an early forerunner of today's mainstream psychiatric practice.

Why had I never heard a word about this pioneer? I regularly wrote about psychological diseases and psychiatric research, and my nose had been buried in psychiatric journals for years. Yet I had never heard anyone allude to Heath in conversation. No one was mentioning his research at conferences. I looked up his name in *A History of Psychiatry*, the historian Edward Shorter's exhaustive examination of the science of the mind from Bedlam to today. Heath wasn't mentioned once in the main text. He showed up only in the reference list, where his name was spelled incorrectly. He had morphed into Robert Heart.

Then I discovered Heath's obituary, published in the *New York Times*. He had died in 1999, at the age of eighty-four, after an unusually long career as a professor and department head at Tulane University in New Orleans. He reigned over Tulane's departments of neurology and psychiatry from 1949 until 1980. The brief piece acknowledged groundbreaking work on schizophrenia, alongside a note that he had been involved in work for the Central Intelligence Agency during the Cold War. At the end, there was a brusque list of his memberships in various golf and hunting clubs and a picture of the man. It was an old photo, reminiscent of a portrait of an actor from Hollywood's golden age: lush, slick, dark hair, with just a touch of gray, and an appealing face with a serious, intense expression. His gaze suggested that he knew something—something essential and important—that his viewer didn't. Otherwise, he was inscrutable.

I found one other picture on the Internet, a black and white so grainy that it must have been photocopied from an old newspaper several times over. It showed a much older man, completely white-haired and wearing a lab smock, attaching a complicated metal device around a human cranium. The apparatus looked eerily like ones that I had seen presented at neurosurgery meetings over the past ten years, as deep brain stimulation—the friendly new name for the brain pacemaker technique—has been introduced to treat a variety of disabling symptoms. Surgical implantation of electrodes in the brain was most commonly being used for people with Parkinson's disease. Today, more than 120,000 patients are walking around with live electrodes to keep their tremors and stiffness in check when drugs are no longer effective.

But I knew that the area where the hottest developments are taking place is in Heath's field of psychiatry. Deep brain stimulation is being described as the field's great new hope. There had been a burst of reports on tests to see how electrodes might treat all sorts of conditions that may be hard to imagine being treated by a "pacemaker"—for example, the obsessive thoughts and compulsive actions we know as OCD, as well as Tourette's syndrome, depression, autism, and anorexia. Even heroin addiction, alcohol abuse, and gluttony have been the subject of electrode therapy. Wherever drugs have stopped working, or never worked at all, patients and doctors are seeking a cure with deep brain stimulation.

The stakes are big. Companies are competing to develop the most discriminating equipment, and market analysts predict that, by 2019, almost $10 billion will be spent globally on deep brain stimulation. Moreover, DARPA, the US Defense Advanced Research Projects Agency, had gotten into the game. In 2013, the agency mostly known for laying the groundwork for the Internet offered $70 million for the development of "next generation" deep brain stimulation systems. These military folks want electronics that will not only stimulate activity in the brain but also continually and instantly read what is going on inside a person's skull. They want equipment that can measure and keep track of a person's brain activity, analyze it, and correct it before the person even acts. According to DARPA's funders, this brave new technology will help heal the thousands of psychologically battered war veterans who return home from combat with post-traumatic stress disorder, or brain injuries caused by repeated exposure to explosions—an innovative electronic cure for the injuries of modern warfare. Changing people's minds in other ways would simply be an interesting by-product.

The story of deep brain stimulation sounds benignly and sympathetically straightforward. Yet, I began to wonder whether there are parallels between what is going on right now and what went on in New Orleans many decades ago. Perhaps, there is even a story that points beyond itself to something deeper. Something bigger . . .

Because, doesn't our reaction to sticking electronics into people's heads actually reflect our relationship to ourselves?

As far back as I can remember, I have been interested in—no, hopelessly obsessed with—understanding how people tick, perhaps, thereby to understand myself better. Whether it is biology, psychology, or psychiatry, I have been interested in what these fields can tell us about ourselves. What they can tell us about our innermost core—or whether we even have one.

What can we, what should we, properly do with a brain? How much should you modify the thing that so substantially constitutes a person, and who decides how far you can go? Is our technology guiding us more than it should? Do we *do* something simply because we *can*? Where is the limit for making people happier? Or calmer? If we can make someone more moral or empathetic, should we?

There is something else at stake for me in this story. Who was this Robert Heath? How can a pioneer who did such work just a few decades ago be unknown in a scientific field that has exploded with news of breakthroughs and applications in mental health that are astounding? I sat with the two pictures—the young dreamer and the old man with the steel apparatus—and wondered.

CHAPTER 1

Singing the Brain Electric

The year 1951 was a little over a month old when the small group of men in lab coats gathered in the operating theater. Anticipation hung heavy in the air, but everyone went out of their way not to act as if this was just like any other day at the office.

The surgeon, a psychiatrist, and the others in turn offered their commentary on the unusually cold New Orleans winter. As they talked, a young woman lay before them on the operating table, conscious but not fully present. For the past six months she had been hospitalized, silent, and withdrawn, and almost immobile. If an aide started brushing her teeth or combing her hair, she would complete the task, but only torpidly.

DIAGNOSIS: Schizophrenic reaction, catatonic type, read her medical record.

The file noted that she was an only child, and had lived all of her twenty-six years with her parents in the countryside. They reported that she had always been a quiet and obedient child. “She’s a good girl,” her mother said.

At the same time, the woman’s health was fragile. There was a long litany of complaints: discomfort, diffuse pain, general fatigue, repeated fainting spells. The year before her hospitalization, these episodes morphed into a chronic condition of irritable confusion and a morbid obsession with her guilt. With tears running down her cheeks, the good girl begged her parents again and again for forgiveness from sins neither of them could recognize—including imaginary sexual transgressions she supposedly committed in her early childhood. Finally, her parents had decided to try a private clinic that promised a definitive treatment for their daughter’s overexcitement: a series of electric shock treatments.

The shock therapy helped a bit, but the woman quickly developed an exaggerated fear of disease. She was particularly anxious that her body was producing insidious, malignant cancers. Eventually, the worries became too much for her, and she attempted to kill herself with her father’s hunting rifle.

A feeble suicidal gesture, the medical file reported, but enough to land the young woman in New Orleans’s Charity Hospital. She had been put in the women’s ward—the white women’s ward—where she had spent the past six months. Then, unfortunately, she had started to experience new delusions and hallucinations. Another round of electroshocks provided relief from these symptoms, but had also thrown her into a state of mute isolation.

There was a separate sheet in the medical record, marked

Patient 4, dedicated to the bold experimental treatment under way on this frigid February evening. That night, a young surgeon named Francisco Garcia was conducting a fourth surgery on the young woman's brain under the supervision of his boss, Robert G. Heath. Four hours earlier, they had opened her skull and cautiously placed a single, thin silver electrode through her right frontal lobe all the way down to the bottom of her brain, leaving its conductive tip in what was called the septum. This was Heath's area: a little region of the brain that he believed served as a focus for emotions, desires, and lust. Heath was convinced the septum was the key to waking his patient from her schizophrenic trance. It was the prince to her Sleeping Beauty.

Her shaved head was covered with a white, caplike bandage. On the right side of her pate, the back end of the fixed electrode jutted out like an antenna. In and of itself, the operation was simple, but Garcia was familiar with how difficult it was to place the electrode precisely. He had to cut through one side of the prefrontal cortex, the brain's tightly folded outer layer, and create a small opening into the cavity of the lateral ventricle, one of two fluid-filled cavities almost at the brain's center. He then had to guide the electrode, following a chain of anatomical markers leading him to the tiny ventricular channel called the foramen of Monro, which was near the brain's midline. Using a technique called pneumoencephalography, Garcia and his helpers would then pump air into the hollows of the brain and take an X-ray to verify that the electrode was in the place they wanted. The process would give the patient a terrible headache later, but there was no way around it.

The electrode now properly placed, Heath took over. He turned toward the technician, a man called Herb Daigle, and told him it was time to connect the electrode to the power supply. Heath then moved gently to the patient's side. She looked petrified, her eyes nearly closed. When he began to speak to her, he adopted an everyday tone, as if the situation were entirely normal. He was here to help; she should follow his directions. It didn't seem as if she had heard anything, but he did not hesitate: He gave the signal to Herb to turn on the juice. The room went completely silent.

They had agreed to start cautiously and stimulate the patient's septum for one minute at 4 volts and 2 milliamps of current. Nothing. Absolutely no reaction. Throughout, they kept an eye out for any sign of seizures, which had been a complication during the surgery for Patient 2 a month earlier. No one wanted to see that again. They watched her blood pressure just as carefully. From animal experiments, they knew that blood pressure could rise sharply when the deep regions of the brain were tickled. But, again, she had no response. The current was turned off. Yet there was no reaction from the young woman.

After waiting for a moment, Heath nodded and Herb restarted the stimulation. This time, the intensity went up a notch, to three milliamps—higher, but still on the low end. The electric pulse was kept on for a minute and a half this time. They watched the patient's blood pressure but her body appeared to give no resistance to the stimulation. They were not sure what this meant. So, quickly, they drew a small amount of blood to test for changes in the woman's stress hormones. Heath once again spoke to the woman, asking trivial, easy questions: *What's your name? Where are you?*

"Hos-pital . . ." she whispered suddenly, and, a moment later, "New . . . Orleans . . . ," this time more hesitantly, almost inaudibly, as if having a voice were new to her, something she must get used to. Heath, who had been standing in front of her until this moment, bent down and laid a hand on her arm. Looking directly into her eyes, which were now open, he asked whether she was in pain anywhere, if she felt anything special.

After a long pause she answered.

"No pain, Doctor."

The surgeon and the psychiatrist glanced at each other above the patient. Heath was glowing. This was exactly what he had predicted, what he had always known *had* to happen. At that moment, right there in the operating room, he might even have allowed himself the luxury of imagining a future in which he finally unlocked the mystery of schizophrenia and found a method to break the hardened shell of the disease and drag the person inside back into the real world.

"Do you see the change?" said Heath to the room, but he was mostly talking to himself. Behind his instruments, Herb was relieved, almost elated. He had helped construct the electrodes in the machine shop, and had spent hours discussing the dimensions and materials with his boss and the neurophysiologists. From the beginning, he had sensed a special quality in Heath, a vision or drive that instilled a passionate desire in everyone around him to deliver their very best. Of course, he was merely a technician, not one of the men confidently assembled around the operating table with their long, prestigious educational pedigrees to back them up. Yet Herb knew he had his role to play in something big, and important.

In total, the little team stimulated the woman's brain six times, at the highest pushing the needle up to 4 milliamps and 8 volts, and holding it there for a good two minutes. She was awake and attentive throughout. She didn't break into long sentences, or elegant argumentation, or even song, but it was the first time in months she had responded to anyone.

Heath made a signal, and finally, the experiment was over. Garcia stepped forward to secure the silver thread sticking out of Patient 4's scalp. They were going to attempt another round of stimulations in a couple of days, and until then, the electrode would have to remain in place. With infinite care, he put a little bone plug into the hole in her scalp, pulled the skin tight around the opening, cleaned the wound, and bandaged it, hoping that would keep everything in place. Inwardly, he crossed his fingers, to ward off the threat of infection. If the site got infected, their work with Patient 4 was effectively done.

Back on the ward, the young woman turned to a nurse and, for the first time in months, spoke on her own initiative. She was tired of all the therapists who had been circling around her bed trying to get her to talk, she complained. She would like to be rid of them.

By the age of thirty-four, Heath had created his own medical fiefdom. He ruled supreme over two floors of the new, boxy building on Canal Street where Tulane University had consolidated its medical research facilities. He had chosen to decorate his territory in the simple, modern, academic style, a classic psychiatrist's couch snug up against the wall, the clean lines and beige tones exuding competence and efficiency. His laboratories were stocked with the most advanced measuring instruments; he had access to a seemingly endless supply of cats and monkeys on which to run animal models; and his operating suite had been customized to his specifications. There was even a room filled with camera equipment, so that every experiment could be documented on film. Most important, he had surrounded himself with top-notch colleagues who were keen to follow him into unknown territory.

In the anteroom of his office, Heath's efficient secretary, Irene Dempesy, was on top of any problem or administrative nonsense that walked in the door, leaving Heath free to envision a different path forward for psychiatry. From his office, he could lean back in his cozy chair and look out on New Orleans's celebrated, and notorious, French Quarter.

When he had told his colleagues at Columbia University about his offer from Tulane, they had reacted with incredulity. They didn't understand how Heath could want to leave a cushy position and great career prospects in New York to go off to such a backwater. After all, he belonged to an urbane milieu that tended to think there was nothing down South but swampland hicks, illiterates, and corrupt politicians.

He understood their objections. He had a lucrative private psychiatric practice on Park Avenue and a tenure-track position at Columbia. New York City was the promised land of American psychiatry. By comparison, Tulane was a one-horse town, backward, without any research tradition. In fact, the place did not have its own psychiatry department. Not yet, anyway.

But he also understood that Tulane was offering him the chance to realize his grand ambitions. The new dean of the medical school, Max Lapham, was something of a tornado, a force of nature that could not be held at bay. Lapham was obsessed with the idea of getting his institution into the big league. He regularly talked of creating a "Harvard of the South," which would require, above all else, recruiting innovative researchers. He needed hungry young men (and a few women) who would grab the money, space, and freedom to pursue new ideas.

During a vacation to Atlantic City, Lapham had bumped into a psychiatrist from Columbia who had mentioned his young colleague, Robert Heath, as one such rising star. Heath was a certified neurologist as well as a psychiatrist. He was known for his hard work in the lab. But what piqued Lapham's interest was the report of Heath's character. The psychiatrist said Heath possessed an unusual dynamism and charisma. He was the type who would get people—including the media—excited about his research.

Lapham decided to get a look at Heath for himself. In Manhattan, he was greeted by a tall, elegant, almost beautiful man who could have easily nailed a screen test for MGM. He exhibited athletic grace and an inner calm, the latter unusual for someone so young. He quickly saw that Heath deserved his reputation as one of those people who naturally captivated a room when he entered it. And he was surprisingly sharp. He was more than current in his knowledge of the brain and the psyche, but at the same time wasn't afraid to argue—convincingly—for his own original ideas. This was the man of Lapham's dreams for the medical school.

Lapham wooed Heath and his wife, Eleanor, to visit New Orleans, but they were barely arrived when the city was hit by a hurricane. The city was shut down for the next several days. Lapham couldn't believe his bad luck. He was sure these two delicate souls from the well-ordered North were ready to decline his hospitable offer to move to the disheveled South. But Heath had just laughed and professed to the dean how he had fallen in love with the place.

As he would later suggest to one of his interns, it was not so much the New Orleans sights and sounds. Instead, as he was being shown around, he had been taken across the walkway connecting the medical school directly to Charity Hospital, and discovered what to him was a veritable gold mine: degenerate, raving syphilitics; catatonic psychotics petrified like awkward, living statues; melancholics who had completely disappeared into themselves and could no longer get out of bed on their own; volatile paranoids who had to be strapped down to keep them from living out their delusions and harming themselves and others. All of them disturbed, tormented souls. And all of them fantastic "clinical material."

Heath's private passion was schizophrenia, the terrible and mysterious disease that struck the young, crippling and darkening the rest of their lives. "It is the most disabling disorder we have," he explained to his medical students during lectures. "Anyone who goes into medicine *must* be interested in schizophrenia."

But what sort of illness was it? No one had found any explanatory changes in the brain structure of schizophrenics, or even biochemical irregularities. In fact, the consensus had formed that it was not an ailment at all but a purely psychological disturbance—a "super-neurosis," in the reassuring words of the psychiatrists. Psychiatry at the cusp of the 1950s was, for all practical purposes, synonymous with psychoanalysis. In the period following the Second World War, Freud's doctrines and a small army of his disciples had swarmed into every psychology department in the United States. Independent psychoanalytic institutes outside the aegis of the university were established left and right, and a private market for the "talking cure" was booming.

Psychoanalysts considered hallucinations and delusions to be psychodynamic imbalances that could be traced back to childhood traumas and, in particular, to bad mothers. They referred to the awful "schizophrenogenic mother," a cold, distant woman who played havoc with the psyche of her defenseless child. Heath thought this was pure nonsense. As though all patients had experienced the same things! "A cat cannot tell you about deviations in its subjective experience, but subjective experiences may very well be the key to understanding psychological illnesses," Heath always argued. "Only people can give us a description of what they experience and what they feel."

Heath had specialized as a neurologist before he had even

formed an interest in psychiatry, and he was convinced that diseases of the mind were caused by faulty functioning of the brain. How else would they manifest themselves? When he had worked as a psychiatrist at the naval hospital in New York during the war, he had fantasized about how dissecting the mental processes would one day explain psychiatry's vague, airy theories about the mind.

So, as he sat down to discuss the terms of his move to Tulane, Heath insisted to his new friend and sponsor Max Lapham that the medical school's new psychiatry department must also include neurology, and he was to be head of both specialties. Heath got his way. In 1949, the year Heath arrived, Tulane was the only university in America to offer training in both disciplines to medical students. Within just a year of setting up his lab, he was exploring the wild idea of placing electrodes into the brains of his patients.

Heath had decided to start with the patients for whom he had the most passion: schizophrenics. The choice was based not just on his personal interest but on a series profound observations too—not least his confrontations with the great psychiatric epiphany of the time, surgical lobotomy.

Between 1936, when Portuguese neurosurgeon António Egas Moniz reported that he had performed the first frontal lobotomy to treat schizophrenia, and 1949, when he received a Nobel Prize, twenty thousand Americans had been lobotomized.

After the war, the demand for the operation had grown sharply, as numerous veterans returned home with severe psychological damage. Nearly 55 percent of veterans' hospital beds were taken up by psychiatric patients. Their relatives demanded effective treatment—no more hydrotherapy or insulin-coma. And it was pressing that these difficult patients had to be released from overcrowded hospitals and sent back home. It's no wonder that doctors were cutting into and lesioning the frontal lobe at such a high clip.

But in some circles, people were worried about this "miracle cure." The results were too mixed. Too many patients were experiencing brain hemorrhages. The procedure was also deemed far too random. Vital areas of the brain were often destroyed unnecessarily.

One of the main skeptics was Fred Mettler, a leading neurologist at Columbia. Uneasy with the number of returning soldiers being referred for lobotomy, he partnered with the Veterans' Administration and the New Jersey State Hospital in Greystone Park to launch the Columbia–Greystone Project. Under the auspices of the project, a team of researchers were charged with testing a more "sparing" procedure, called a topectomy, which removed much smaller portions of the frontal lobes. As it was the first study of its kind, the researchers took a disciplined, scientific approach, selecting their research subjects so that they constituted a homogenous group that could be compared.

Mettler tapped Robert Heath to serve as the senior psychiatrist on the prestigious project. His chief responsibility was investigating and assessing the mental condition of the patients before the topectomy and up to six months afterward. There were over five thousand patients at Greystone Hospital, and forty-eight people with either schizophrenia or depressive disorders were chosen as experimental subjects for the project. Half of them were control subjects who received no treatment, while the rest had a small portion of their frontal lobes removed.

As Heath pored over the findings after many months of conversations, operations, and investigations, he saw that the method was not working—at least, not as the team had hoped. Only nine of the twenty-four patients who received a topectomy achieved any lasting improvement, and the procedure had side effects, not least a flattening of emotions. Moreover, three of the patients improved for a short time only; the other twelve showed no change whatsoever. Heath, who never cared for lobotomies, reported to Mettler that the topectomy was no better.

Yet as he worked through the many assessments, he also realized something deeper: If they could not stop schizophrenic symptoms by cutting into the cerebral cortex, this coiled, outermost layer of the brain, the source of the condition must lie elsewhere. Perhaps, he thought, regions below the cortex were also functioning abnormally and the key to getting the whole system in line could be found there. With this in mind, he looked at the charts of the patients for whom the topectomy was most effective in curbing negative feelings. They all had removed regions of the cerebral cortex that were connected to the same area deep inside the brain, namely the septum.

This thinking meshed with a crucial theoretical inspiration that Heath would encounter when he returned to the ivory towers of Columbia. The Hungarian psychoanalyst Sandor Rado, head of the university's psychoanalytic program and a student of Freud himself, was urging his colleagues to investigate the biological reality of psychological states. Rado's mentor had abandoned the project long ago because it was too difficult, but Rado believed that science was now ready for it.

For Rado, a stocky little man with a strong accent, the very core of schizophrenia was not the hallucinations or delusions of being Napoleon, the risen Messiah, or the helpless victim of harmful rays beamed down from Alpha Centauri. No, the fundamental symptom was an absence of positive feelings, what Rado called anhedonia. Schizophrenics were filled with the negative—anger, anxiety, despair—and reported feeling incapable of experiencing any form of pleasure. They simply did not know either joy or the feeling of happiness and, therefore, had lost the ability to regulate their behavior normally. If humans had two inner principles to guide their conduct—negative and positive feelings—then schizophrenics were handicapped because they were tethered by only one. It was like being a horse that was constantly whipped no matter what it did, never getting a carrot. It can't help but move in zigzags and circles. There is no longer any logical direction.

Heath knew that Rado was right, and further, he had arrived at an idea for how to investigate the physiology of psychology. Whereas traditional psychosurgery destroyed parts of the brain forever, he wanted to develop a treatment by investigating what happened to a patient when different regions of the brain were manipulated *temporarily.* And he planned to use the brain's own driving force, namely its crackling electrical activity. In the case of schizophrenia, he imagined that you might be able to force an anhedonic brain to feel enjoyment and pleasure by stimulating the area where these emotions resided. Such an emotional revelation might just be able to rip the afflicted person out of her anxiety-filled shadow world. Heath would prove to the patient that there was a carrot. And once she could see there was a direction forward, that there was light ahead, she would be able to engage with reality and find her own way.

It was worth a shot. A magnificent shot.

At Tulane, Heath gathered around him a loyal guard, a group of musketeers, people representing different professional disciplines but sharing his overall vision. A phalanx of young, hopeful psychiatrists and a few neurologists followed him down from New York, in particular his good friend, the psychiatrist Russell Monroe. In addition, there were Robert Hodes and Walter Mickle, both neurophysiologists Heath met during the Greystone Project; an experimental psychologist, H. E. King; an electrical engineer, Hal Becker; and, to cover all the bases, a psychoanalyst, Harold Lief.

Heath also handpicked his own neurosurgeon—Francisco Garcia—to join him from the Greystone Project. Unfortunately, this was not entirely a choice of his own making; Tulane's surgeons were not too keen to sign up. The type of operation Heath was suggesting had never been attempted before, and many responded that they could not see what good would come from it. Some even replied that it was not wise to rummage around in the brains of the mentally ill on the basis of wild theories and presumptions. It was too risky, for the patient, of course, but also for any doctor who agreed to transgress the conventions of the field.

Heath did not give a damn about conventions. He loathed being bound by "business as usual." But he now could not escape the obvious fact that he would be talked about at the university, and that there would be some colleagues ready to scupper him. Perhaps they envied the special position that Lapham had carved out for him. Regardless, he had to fight these naysayers. They were just old geezers who did not understand that new paths had to be blazed because the old ones were muddy and full of potholes. Heath could be charming, but he could also be tough as nails, and he quickly earned another reputation, this one for his explosive temperament.

Heath had barely put down stakes in the medical building when he turned his attention to reforming the decrepit Charity Hospital, his gold mine of experimental subjects. The hospital was still run by Catholic nuns, and they were resisting the changing order. To the extent the patients received any treatment, the nuns' methods were outdated, to say the least. Heath discovered that syphilis patients were treated with an obsolete malaria therapy. The nuns actually summoned John Walker—called "Jungle Johnny"—from the university's tropical medicine department to come prancing over with a vial of malaria-infested mosquitoes to bite the patients in the hope that a high fever from an infection would purge the syphilis bacteria from their system. It would have been laughable, if it weren't so horrible.

Just as bad, the psychiatric ward on the third floor was like a big cage. Most of the patients seemed to have been admitted merely for long-term storage, and, in practice, lived their lives tied to their beds. They could not even be examined without calling personnel to loosen their belts and straitjackets. Heath, as newly minted head of psychiatry, complained in writing to the hospital management. At a subsequent meeting with the hospital's psychiatric committee, he fought to get room for more beds and to have a private room available for every third patient.

His challenges against the status quo were too much for some of New Orleans's old guard. Felix Planche, a veteran of Charity Hospital for many years, was incensed that the insane were being marked for special treatment. Ordinary medical and surgical patients were perfectly happy in their open wards, he said. And, he argued, the hospital was a charitable institution, offering free care to patients who were the lowest of the low in the city. These were people who were not accustomed to luxury; indeed, they would be uncomfortable in it. Heath's reforms were sheer financial folly, Planche excitedly asserted. The proposal would cost $400,000!

Heath stood his ground. The proposals made excellent economic sense, he replied, if you want to see improvements in your patients, get them out of their sick beds, and, ultimately, prevent them from ending up as chronic cases—either at Charity Hospital or the large state hospitals scattered around Louisiana. He aimed to isolate the most violent patients in private rooms so they would not disturb the progress of other patients and the fragile process of a cure. The third floor—*his* third floor—was to be rebuilt as model for the whole country, a ward that not only treated patients effectively but also attracted the best minds in the field of psychiatry.

Heath won the argument, and the major reconstruction of Charity Hospital was completed in 1952, the same year that he first presented to colleagues from outside Tulane his novel work in stimulating schizophrenics. As lord of his green plantation, he wanted to show his former bosses and mentors at Columbia what he had accomplished.

With a small group of colleagues, he founded a new professional association—the Society for Biological Psychiatry—which he envisioned as the force behind a major movement to discover better treatments for psychological illnesses. The group was comprised of the young and ambitious, and the society was far more than an extracurricular hobby to them.

At the lab, it became the rule rather than the exception that people met to conduct experiments at night, after everyone had finished their clinical duties. Using animals, they practiced surgical techniques and developed logical and scientific bases for their work with human patients. The septa of countless cats—a favorite experimental animal for the nascent field of neuroscience—were surgically removed, to see what would happen. The cats showed biochemical disturbances and behavior resembling something like schizophrenia. Most of the animals became lethargic—sometimes even catatonic—and their stress reactions and hormone levels went off the charts. A few others became aggressive and unpredictable—classic symptoms of paranoia. Just as Heath predicted, the results pointed to the septum playing a decisive role in a range of schizophrenic symptoms.

There were other signs that Heath's intuitions were correct. A handful of Rhesus monkeys with electrodes implanted in their craniums were regularly stimulated in the septum. The reactions were always the same: The small primates livened up with each stimulation, and even seemed to appreciate the treatment. Astonishingly, the monkeys quickly learned to work the contact on their own, pressing it over and over, as long as the electricity was on.

Then there was the pièce de résistance—the twenty-two human patients whom Heath and his surgeon had operated on and stimulated. The team had studied some of them for months after the operation. The wealth of data practically burned their hands, it was so eager to be presented to the world.

Heath decided to call in the big guns of psychiatry for a symposium, entitled "Studies in Schizophrenia," which would be held in New Orleans. On his home turf. In the front row was Sandor Rado, the godfather for Heath's whole enterprise, and Fred Mettler, the neurologist who trusted him to find something better than the crude option of lobotomy. Mettler not only was Heath's boss at Greystone but had also taught Heath how to do electrode studies on monkeys. The atmosphere was tense. With Mettler especially in mind, Heath could not help but view the presentation as a sort of "graduate exam." Over two days, the guests listened as the team from Tulane presented their findings.

"It is not our intention to announce a new treatment," Heath insisted when he took the podium to open the proceedings. "We have obtained some therapeutic effects, but much work remains to be done in this regard."

The lecture from psychologist H. E. King garnered special attention from the assembled dignitaries. King had tested nine patients before stimulation of their septa, and again four months later. All the patients had gone through an impressive battery of psychological tests to map and quantify their intelligence, personality, memory, motor skills, and of course their schizophrenic symptoms. Case by case, King walked the audience through what he had seen and measured. His conclusions were mixed.

Patients 2 and 12 were significantly better after treatment, and King chose to dwell on Patient 2 in particular. She was young, just seventeen, and had presented with schizophrenic symptoms two years before; she had been hospitalized for a year. Before the operation, she had been catatonic as well as anorexic, virtually starved. The operation itself had been nothing to brag about. After being stimulated on the operating table, the fragile woman went into seizures that lasted several minutes. The convulsions continued the next day, and the patient remained in a comalike condition for two days after that. But then they saw progress.

The girl acquired an interest in her surroundings, began to speak and eat, and, after two months, according to her family, was "friendly, outgoing, spontaneous, and cheerful—in fact more so than she had ever been in her life." Four months after the operation, she was sent home and resumed her education.

And now, over a year and a half later, she was still progressing. She was no longer anorexic. She was doing well academically. She went out with boys. She even avoided breaking down when, shortly before the conference, she lost her beloved sister.

"Nothing more than the usual grief reaction," King said with obvious satisfaction.

Then King cleared his throat and moved on to the other cases, whose results were, sadly, not as spectacular. For Patients 13 and 17, there was a marked improvement, but the condition of four other patients remained unchanged, and a single patient, number 8, had actually experienced an exacerbation of his illness. The team did not have a good explanation for why some patients had responded to the treatments and others hadn't.

Yet, he said, by way of conclusion, it was true that the procedure *could* work, and that more research was needed.

As an extra treat, there were findings from some of the controls, patients without schizophrenia who were operated on for comparison. Heath and Garcia had decided to stimulate a few pain patients, for instance. They had wanted to test the idea that a feeling of intense pleasure in itself can block the feeling of discomfort, simply because the two mutually preclude each other. A brain operation might not seem the obvious thing to do in a serious, chronic pain case, but there had been some limited success in dampening pain through lobotomy, and the Tulane patients themselves were willing to undergo the procedure. They were willing to try anything.

One of the patients was a severely arthritic man. Two terminal cancer patients (one woman with uterine cancer, one with breast cancer) had also volunteered. For all three, medicine could no longer soften their pain. Amazingly, a few stimulations of their septa curbed it quite effectively. One of the women had lived for seven months after her operation, and a single stimulation session of fifteen minutes had freed her from pain for an entire week. But the effect reported by these patients went well beyond pain relief. While their brains were being stimulated, the patients perked up and grew more alert. Throughout, their general well-being improved.

Of course, Heath said, these were only preliminary findings, but given their dramatic nature, he thought they were worth sharing. He recognized that a human being's thoughts and feelings—"their internal milieu"—influenced the diseases of the body. It was an idea to pursue.

The wealth of data, the films of the patients before and after stimulation, and especially the theories behind it all were a lot for the visitors to swallow. On the final day of the symposium, it was finally their turn; one by one, they presented their comments on the research in short, impeccably crafted speeches. Most began with praise for the enthusiasm displayed by the Tulane group—it was wonderful to see so much initiative on display. But then they had to raise some reservations. Weren't their ideas a bit *flighty*? Couldn't they modify the operation to go easier on the patient and lower the risk of hemorrhage and infection? And honestly, didn't the experiments lack proper controls?

Psychiatrist Herbert Gaskill from Indiana University soberly noted that an improvement in half the schizophrenic patients did not say that much. Some earlier postulated "cures" for the illness appeared to have the same effect at the beginning, but if a doctor waited five years, the patients often ended up back exactly where they had started. George Ham from the University of North Carolina was equally skeptical. Heath and his people had definitely hit on some interesting findings—especially when it came to the patients' biochemistry and hormonal changes—but it was hard to maintain that these were due to electrode stimulation. Wasn't it the case, asked Ham, that patients received "an intense interest and attention during the post-operative stimulating phase"? Wasn't the procedure itself an "extreme black magic," with a rather intense "narcissistic reward to the individual for improvement"?

In other words, might this all be due in large part to the power of suggestion? A placebo effect?

Finally, the distinguished Fred Mettler pulled out his critical knife and cut to the bone. His eager, young research colleagues might be engaged in wishful thinking, he more than hinted. In some of them, he sensed an almost evangelical faith in the project. Their data and observations might even have been tailored to confirm the theories that Heath and his group so wanted to prove.

This hung in the air for a moment. Then Mettler closed the discussion: "In Dr. Heath, a former student of mine, I do indeed take pride even though (let it not be forgotten), I do not agree with him."

The farewell dinner that evening was excruciating. It had been arranged as a celebration at one of the French Quarter's most renowned eateries, Antoine's, but it was far too late to cancel the reservation in light of the sour mood. The high notes from a jazz band playing in the street could be heard as Heath led his guests to the restaurant. Fortunately, there was so much noise inside that the taciturnity among the researchers was hidden.

The Tulane folks were shaken. They could not believe the reception they had received. What was supposed to have been their entrance onto the big stage was looking more like the final curtain. A few of the most temperamental were so furious at the reaction from "these old sticks" that they made their apologies and stayed home. Heath knew that, later, it would be up to him to get the team back on track. He would have to explain to them, promise them, that all their enthusiasm, the many nights of experiments and discussions, had not been wasted. That, over the long haul, their plans would be realized. Their way of doing things would be accepted.

That evening, as Heath sat at the far end of the table, it almost hurt to keep the façade of a smile on his face. He looked at each of his guests in turn, struggling to understand why and how they had been so unwilling to see. He was well aware there would be resistance and doubt along the way. The modern never won over the established without a fight. But this cold, blanket rejection had hit him hard. Was this what he was going to face from now on . . . and for how long?

CHAPTER 2

The Second Coming

That's not true."

I had whispered this spontaneously to the person sitting next to me in one of the back rows of the large, dark conference hall. The glance I received in response did not invite elaboration. At the speaker's podium, a Dutch neurosurgeon had just said that attempts to treat psychiatric illness with deep brain stimulation "began in 1999."

"Try half a century earlier," I next muttered, so softly that the man next to me could not hear. Still, I couldn't help myself from making the point out loud. The speaker had gone on to reference two "innovative and groundbreaking studies" also published in 1999, where, within a few months of each other, a Belgian and a Dutch research group separately shared results of experiments implanting electrodes in patients suffering from compulsive behavior and Tourette's syndrome. Both studies involved only a handful of people, but the deep brain stimulation did seem to mitigate their symptoms. It was enough to open the floodgates for further research.

"Since then," the speaker continued, "we have seen the method spread like a wildfire in psychiatry. Today, there is pretty much no psychological ailment or condition for which an attempt is not being made to treat it with electrodes."

News of these experiments had brought me to Maastricht, where nearly seven hundred neurosurgeons from virtually everywhere in the world had gathered for three days to talk shop in the fall of 2015. Most of the attendees were men, clad in identical charcoal-gray suits that seemed to have all been bought from the same store. Boisterous chatter resonated throughout the conference center, a space so anonymous that if I closed my eyes for a moment, I completely forgot what it looked like.

I felt like a stranger. For years as a journalist, I had grown used to mixing with psychiatrists, psychologists, and geneticists. I was familiar with their jargon, the way they spoke about the psyche. Here I was surrounded by people blathering about technology. Their eyes lit up anytime someone started talking about some new style of electrode that could channel current in various directions, or smart rechargeable batteries that lasted ten years inside a patient's body. During the breaks between lectures, I watched as they lovingly caressed the advanced surgical hardware on display—gadgets that, to the untrained eye, looked like modern instruments of torture.

Deep brain stimulation was now the cutting edge of neurosurgery; it was where the most ambitious surgeons wanted to be working. These days, it seemed wrong to squander manual dexterity on the traditional work of removing tumors or repairing ruptured aneurysms, and it granted no prestige whatsoever. Operating on the psyche—that was the future. The neurosurgeon was an artisan of the mind.

But something bothered me. Notable by their absence from this rapidly burgeoning field were psychiatrists. I had estimated from the literature that experiments using deep brain stimulation had been conducted for thirteen different psychiatric diagnoses, and that nearly a hundred clinical studies had been published from around the world. But in Maastricht, the neurosurgeons told me they had been unable to tempt any psychiatrists to join them. There were probably only a couple dozen psychiatrists in total who were directly involved with brain stimulation experiments. When surgeons applied to present their research findings at psychiatric conferences, they received polite rejections. There was simply no interest.

I had come to hear the current buzz in the field and compare it to the history of Heath's lab that I had been unearthing. Each time I addressed one of the gray suits, I brought up Heath's name. Almost all reacted with the identical blank expression. Maybe they had bought it at the same shop as their suits. "Never heard of him" was the typical response. The surgeons had their own creation story for deep brain stimulation. This story went that the method was discovered in 1987 by the French neurosurgeon Alim-Louis Benabid—a man who had since received many awards and huge recognition for his achievements. Everyone could recount the story of how Benabid's breakthrough was the result of a happy accident.

Benabid was in his operating theater in Grenoble, preparing

to treat a Parkinson's patient. A small lesion was to be created in the area of the brain known as the thalamus, which was expected to still the patient's tremors. The operation was routine. Benabid, as usual, had placed an electrode in the patient's brain to cauterize the tissue in the right place. But that day he decided to experiment a little. He stimulated the patient with different frequencies from the one he usually used. When he turned it up to 100 Hz, the man suddenly stopped shaking. Fast-forward to the year 2000, and after almost two hundred clinical experiments, high-frequency electrode stimulation was approved by the US Food and Drug Administration for the treatment of Parkinson's and the movement disorder known as "essential tremor." Today, the treatment is available everywhere.

Meanwhile, there is no official approval for using it to treat any psychiatric conditions. Psychiatrists are caught in an experimental limbo, where permission has to be sought on a case-by-case basis. For some diagnoses—for example, obsessive-compulsive disorder (OCD)—results have gradually accumulated from many one-off experiments. For others, research has been entirely haphazard, and studies have been small—five patients in one place, three in another; there were even several case reports with only a single patient as the guinea pig.

One of these reports dated from 2012. In it, a German surgeon named Volker Sturm described how he had implanted two electrodes in the amygdala of a thirteen-year-old autistic boy who had serious problems with self-harming behavior. The behavior stopped, Sturm claimed. Sturm has since retired. Nowhere can you find additional references to the case. There is no way to know whether others took over the research, or if attempts were made to disseminate the method—or even if there has been any follow-up with the boy to see if the effect had lasted.

As I walked around the conference center, I sensed a ghost hovering above my shoulder. Heath's story, his old research papers, his long-since-dead patients identified only by number, were the sounding board for everything I was hearing. Parallels could be drawn between Heath's era and the current one, and the comparisons were striking—even eerie. Among other things, Heath had been criticized for relying on half-baked theories to support his method of applying electricity to his patients' septa, but what were the rationales today? What were they based on?

I noticed that the use of stimulation in the treatment of depression was very close to Heath's original approach, though he was never mentioned. Even now, it was described as treating "anhedonia," the inability to experience joy, by firing up the brain's reward system. Many of the neurosurgeons were going directly after the little nucleus accumbens, the area of the septum that Heath may well have been hitting in many of his experiments.

Other "targets," as the surgeons called the stimulated areas, were chosen on the basis of old psychosurgical operations. This was the case, for example, in the two 1999 experiments involving OCD and Tourette's. Back in the 1970s, people had known how to treat severe instances of both ailments by creating discrete lesions to the brain. You could go in and remove a little bit of two particular structures in the thalamus to mitigate the more violent behavior in certain patients. It was precisely in these areas that the Dutch and Belgian surgeons had placed their electrodes.

There were also examples of pure coincidence. For example, observations of OCD patients had inspired surgeons to place electrodes in a completely different group of people—namely, alcoholics. The idea arose because some patients who were already in electrode treatment for their compulsive behavior experienced the "side effect" of getting rid of their alcohol abuse. They simply reported less desire to drink. Research groups quickly sought alcoholics interested in taking part in an experiment. Others thought that if you could stimulate your way out of alcoholism using this technique, then why not try the same thing with heroin abuse? Abuse was abuse, with different substances drawing on the same brain mechanisms—or so was the thinking.

At the time of the conference, drug and alcohol addiction were listed as maladies in the American Psychiatric Association's *Diagnostic and Statistical Manual*, but compulsive overeating was not. Nevertheless, some psychiatrists were claiming that gluttony was also a form of addiction, and many surgeons had received permission to do deep brain stimulation on selected patients who were morbidly obese. The rationale was that areas of the hypothalamus, which regulates appetite and the desire for food, were well mapped. The husband-and-wife surgical team of Alessandra Gorgulho and Antonio De Salles from UCLA summed it up in a 2014 paper: "As neurosurgeons have direct access to the brain areas controlling food intake, it is natural that their expert knowledge of stereotactic techniques should be used to reach these centers and modulate them." That was a firm statement. Their first six patients had already been treated in São Paulo and were being monitored.

In their paper, Gorgulho and De Salles argued that once you started to see good results with obesity, anorexia was an obvious next candidate. There was no effective medical treatment for the condition, and psychotherapy helped a small minority. Half of anorexic patients became chronic hunger victims; their suicide rate was alarming. Four research groups, flung from Shanghai to Toronto, had already reported that electrode stimulation helped upward of half of severely anorexic patients to eat and gain weight.

This all appeared self-evident among the gray suits and professional PowerPoint presentations. *Of course* you can treat gluttons, heroin addicts, and recalcitrant anorexics with a well-placed electrode in the brain. Because that is where the problems lie, right?

Yes, but only because something decisive has changed in how doctors and the public at large think about the mind and mental health.

Over the past few decades, the psyche has been compressed from a slightly airy, fluttery concept to a material thing solidly anchored in the gray matter of the brain. The ailments of the mind that we previously believed were due to circumstances, life conditions, and other external factors have been pushed inside the brain's tissues and must now be treated there. On top of this upheaval came a shift in the very way science itself views the brain. The idea of the brain as a chemical soup was ditched in favor of the brain being seen as a mutually connected network of electric circuits. And whereas people once talked about neurological and psychiatric diseases as an expression of imbalances in the brain's neurotransmitters—substances such as dopamine or serotonin—today we increasingly describe these conditions as "circuit pathologies." This is because it is now clear that the electrical network activity to which the neurotransmitters give rise is disturbed in these diseases—it is too high, too low, or perhaps irregular. I heard this new set of explanations articulated from the speaker's podium in Maastricht: "We can study individual patients when we turn the stimulation on and off, and we are moving toward an understanding of how specific behavior is connected to specific circuits and synapses."

Electrodes have something else going for them: They are far more precise instruments than drugs. The psychoactive drugs patients gulp down are typically directed toward specific receptors on the surface of brain cells—receptors that give rise to a signal when they are struck by a particular hormone or signaling molecule. But the cells carrying these various receptors are often spread out across most of the brain, and a drug does not distinguish one area of the brain from another. The drugs have to flow freely to have an effect. The cerebral cortex, the thalamus, or something else entirely—it does not matter where they find a receptor to attach to.

An electrical influence is fundamentally different, because each electrode affects only the tissue immediately surrounding it. So you can target the network of cells whose activity is problematic and correct how it works.

• • •

That was the theory, at any rate. I quickly realized that what actually happened in the brain tissue was still quite hazy—not just to the neurosurgeons but to everyone, in fact. Even with respect to Parkinson's patients, tens of thousands of whom had been operated on, the surgeons did not know what the mechanism in deep brain stimulation was. The experts disagreed among themselves on the extent to which the charge to an electrode stimulated or inhibited the area immediately around it. It looked as if a high frequency inhibited, whereas a low frequency stimulated. In other words, one could turn the frequency up or down and get different effects from the same electrode. The effect also depended on whether one stuck the electrode in areas that were full of cell bodies—the proverbial gray matter—or in the white matter, the main cable connections of the brain, consisting of cell axons bundled up in large fiber tracts.

There didn't seem to be much knowledge about what was going on at the microscopic level. Nor was there much clarity about the overall neural circuits. There was no agreement about where best to place the electrodes. Different groups around the globe placed their electrodes in different areas of the brain, even though they were treating the same type of patient. There was even a joke about it: It was not the patient's disease that determined which area of the brain was stimulated but the patient's zip code. As the Canadian superstar surgeon Andres Lozano from the University of Toronto acknowledged: "The field is characterized by a Wild West mentality, and no brain cell is safe."

Lozano was tall and slouching, a chronic expression of

concern cast across his face. His expression darkened even more when he counted up the areas of the brain—forty-two—that were used as targets for stimulation as of that day. There were ten areas being used to treat depression, and six for OCD. Even more striking, the same area was often being used for completely different diseases.

"We need to coordinate and find out what actually works best," said Lozano to the packed hall, flashing a PowerPoint slide that reminded me of Heath's brain maps. Lozano showed that the most popular target among his colleagues was the nucleus accumbens. This tiny area was being stimulated in depressive patients as well as those suffering from either anorexia or morbid obesity, just as it was being used to treat OCD, alcohol abuse, and heroin addiction. The nucleus accumbens was smack in the middle of what Heath called the septum.

That seemed more than a coincidence. But when it comes down to it, we know very little about mental illnesses. If you speak long enough with psychiatrists, they will usually admit that their field has advanced only to about where the rest of medicine was a hundred years ago. In reality, the best you can do is address general symptoms—be they depressive episodes or hallucinations. They have no biological mechanisms on which to pin their patients' troubles.

Illness, sickness, disease. Of course, diagnoses were the main reason we were sitting in a dark hall in a boring concrete building listening to speech after speech, but driving my private fascination with electrical brain stimulation was something completely different. I was more interested in the fact that manipulation of the brain is a manipulation of the innermost self—the *me* that is found somewhere within that pinkish clump with its hundred billion connected cells. This remarkable technology raises that deepest of questions: Who am *I*?

As long as I can remember, this has been the core question to which I return again and again. Maybe it goes back to my childhood. My parents were not a good match, and they both wanted the other to behave differently. There was a lot of talk about changing, and my father especially believed that you *could* become someone else—"it is a question of rearranging bits of information in your head," he asserted. But I could see that he actually changed very little during his life. Even though he wanted to, and the topic was argued this way and that until the whole family was blue in the face, he just couldn't manage it.

I too was well familiar with the desire to become someone else. Or at least another version of myself—better, somehow. As a journalist, I had been chasing after studies that map the plasticity of the mind. But speaking to psychologists and behavior geneticists, I had learned how our genes draw the contours of our personality and that personality tends to be fairly stable. "Come to terms with yourself" seemed to be the message.

But genes create personality by fashioning the way the brain reacts to external stimuli. If you go the other way—fiddle directly with the brain—personality must follow along and change. And this is precisely what electrical brain stimulation has been proving.

Take the story of Patient B. After suffering from OCD all his life, this fifty-nine-year-old Dutchman was successfully treated with deep brain stimulation in his nucleus accumbens. Everything went smoothly, and the symptoms disappeared. Then, after six months, an odd side effect appeared: he developed an unexplainable appreciation for Johnny Cash. Up until that time, Patient B had had a fixed musical taste that ran in the direction of Dutch rock and roll and the Rolling Stones. Country-and-western music left him stone cold. Now, he craved hearing the Man in Black, and said he really identified with Cash's songs.

Patient B experienced a profound change in himself but, at the same time, he could switch it on and off. His passion for Johnny Cash diminished as soon as the stimulator's battery ran low or the cord was pulled.

This on/off phenomenon applies not just to musical tastes but to personality traits too. The group at the Academisch Medisch Centrum in Amsterdam, which operated on Patient B, described how two other OCD patients were transformed by their stimulator. As soon as the current got a little too high, both patients reported an inflated feeling of self-confidence and a previously unseen impulsiveness.

The chief surgeon on the operations was Richard Schuurman. A tall, lanky Dutchman, he was the only person at the conference in Maastricht who mentioned personality. He began by describing a Dutchwoman who, like Patient B, was liberated from her severe OCD after deep brain stimulation. But she did not care for the price to be paid because she also lost her powerful predilection for perfectionism. It was one of the symptoms of her condition, but she felt her personality was too slack without it. She missed her disease.

For another woman, it was the reverse. The electrical stimulation turned her from a reticent, introverted person into a bubbly extrovert. As a result, she developed a problem with alcohol and difficulties in her marriage. But she apparently did not care. She said she loved her new personality.

At one point, Schuurman asked the hall, "What is the self?" He wasn't expecting an answer, but instead indicating that his observations and those of his colleagues were raising questions that challenge our habitual thinking. We have thought of the self, or our "ego," as some sort of inner core. Beneath all the possible layers of upbringing and culture, there must be a genuine or true self that we can identify if we go to the trouble of paying close attention. If your life gets screwed up, you simply need to "find yourself," as the saying goes.

The concept of this stable inner core is ancient and tenacious, but it is an illusion. It fits poorly with what science has discovered. No one can point to any special area of the brain or any particular neurons that create either personality or a sense of self. In fact, the pallid-looking Schuurman could have just come right out and said it: *The self is what he and his colleagues make it into.* Result after result was clear: No fixed inner self existed in any of the patients. Rather, you are the condition of your brain at any given time. Apply a bit of voltage here or there, and you become someone else.

This realization has major consequences, none of which were discussed within my hearing at the conference. But if we are to recognize that the self is fluid and malleable, and decide to explore how we might shape it with technology, we have to think about the question of boundaries. Is there a limit to what we can imagine changing through electrical stimulation, and will any line be drawn in the future?

We could take psychopaths as an example. Researchers are studying their brains for physiological defects—and very slowly beginning to identify them. What if you could actually put an electrode in the head of an otherwise unfeeling human being and make that person empathetic? Would that offend our moral intuitions, or would we think it was an excellent treatment?

In 2012, the Italian neurologist Alberto Priori argued that we should use our knowledge of the anatomy of morality—how the brain processes moral questions and makes moral assessments—to develop "therapeutic strategies for neurologically-based abnormal moral behavior." Among other things, he suggested that we use deep brain stimulation to treat aggression and "other forms of pathological antisocial behavior, including violent sexual criminals and pedophiles."

Priori's words are strikingly reminiscent of Heath's treatment of the young homosexual man, Patient B-19, in 1972—just with a new epoch's definition of the "undesirable behavior" that medical science would like to remedy. Although the electrodes are thinner and the technology more sophisticated, today's work in deep brain stimulation has not evolved much since Heath's time. Modern surgeons have scanners that allow them, in principle, to see exactly what they were doing. And, of course, they also know more about the brain than scientists did in the 1950s and '60s. But the techniques and methods are pretty much the same.

In contrast to that earlier era, however, there is enthusiasm for the research. It bounced off everyone in the conference center in Maastricht, this sense that they held the technology of the future in their hands. The willingness to take a risk was there again. The researchers involved were feeling their way forward—sometimes one patient at a time. A surgeon or perhaps a psychiatrist got a bright idea and sold it to a patient, interested collaborators, relevant academic boards. Wouldn't it be logical to stimulate here and there? And anyway, the treatment was reversible. You could just pull out the electrode if things did not pan out.

But even today, operating on brains does not come without risk. According to some studies, up to 3 percent of patients experience serious hemorrhaging, which can result in brain damage. Nearly 5 percent get an infection. With those sorts of odds, good arguments need to be on the table before you resort to the knife. The top argument almost always boils down to the same thing: *We are talking about severe cases, people who have been given up on and have no other choice.* As a young surgeon told me in Maastricht, parroting the line, "When you are dealing with patients who are suffering and have tried everything that is out there, don't we have a duty to do something if there is even a possibility it will help?"

It is a good question. And almost word for word what he would have read in Heath's original articles from the 1950s, if he could be bothered to look them up.

Only a few of the neurosurgeons at the conference were familiar with Heath, but among them was the Canadian superstar Andres Lozano. When I asked his opinion, he did not hesitate in calling Heath "a pioneer." But in the same breath, he hurried to stress that his predecessor's experiments were "unethical even by the standards of the day."

Lozano related how patients in the 1950s and '60s were not provided with sufficient information to be able to give truly informed consent—the consent we today take to be the absolute precondition for medical experiments and treatments. I was about to object that the notion of informed consent was quite different in 1950, when faith in authority was strong and the ethical debate was pretty much nonexistent. But before I could do so, Lozano admitted that the field today has been guilty of close to the same offense. His own patients were presented with a ten-page-long form with information about the procedure and its possible complications, but, as Lozano said, the text was, for all practical purposes, incomprehensible to them. It was full of abstruse formulations from university ethics boards, put there so that they could wash their hands and say they had done their duty. Ordinary patients had no way of understanding what they were getting themselves into.

I wondered if Lozano could hear himself ranting. If his comparison held and contemporary researchers still believed they were on the right path, then why was he so fast to condemn Heath? Was it because the results today were better?

Certainly not always. Typically only a third to a half of patients show any effect from the treatment—far from a complete cure, but still a measurable effect. There are also experiments in which there is no effect at all. Moreover, many of the speakers in Maastricht indicated that there is a lack of follow-up on psychiatric patients after brain stimulation. They disappear from view, and disappear from the scientific literature. You might get a situation report at some point shortly after the operation, but would have no idea what happened over a longer stretch of time.

These were all sins I had seen Heath accused of and condemned for. The many parallels between him and his successors made me wonder what exactly had gone wrong for him. It was clear that the answers were not in the research or the academic papers I had slogged through. I needed to find other ways into his story to understand it. I had no idea how difficult this would prove to be.

"Materials on Heath's work are not accessible to anyone but researchers pursuing similar projects."

So succinct and categorical was the answer I received from the head of psychiatry at Tulane University that it was like getting doused with a bucket of cold water. I didn't understand. Dan Winstead actually held the Robert G. Heath Chair of Psychiatry, but he apparently had no interest in talking about his predecessor. In a follow-up e-mail, I asked whether a "serious" writer might gain an audience?

"That has not been our policy."

I thought it would be easy: that I would simply write to this respected institution—where, after all, Heath had been the king of psychiatry and neurology for three decades—and they would welcome me in and open the archives. If the man had been unjustly besmirched and forgotten, you would think Tulane would be interested in someone taking a fresh look at the story. But no.

"The decision was made by our legal department, and they have recently confirmed this policy."

Our legal department? What did they have buried in their archives?

A direct inquiry to the university library did not reveal much more. At first, the librarian was not directly discouraging. After a little bit, she admitted that the university had no real records with respect to the late Dr. Heath.

Huh? Academics usually left their academic and personal papers to their university. After several more inquiries, I was told that Tulane had a digitized archive. In exchange for $20, the librarian would send me a DVD. The DVD had copies of old press releases, newspaper cuttings, and a single filmed interview with Heath, made after his retirement.

It wasn't much.

My next option was to find people out there—firsthand sources—who had worked with Heath in his laboratory, maybe even attended to his patients, witnesses who could talk about their experiences and tell me about him.

A Robert G. Heath Society had been founded by former students. Every so often they met in New Orleans to honor him. On their home page, there were pictures from their most recent reunion: one featured a large decorated cake shaped like a brain, the whole thing swaddled in a brownish marzipan. The participants were mostly gray-haired men in suits that no longer fit. The society's most recent newsletter was several years old, but I tried e-mailing a few members of the board and received an answer from one of them.

Robert Begtrup is now a child psychiatrist in Memphis. He was a little surprised to hear about the cold shoulder Tulane had given me, but believed it must be connected with the anger that exists "out there" and which the society experiences now and again. As chairperson, he had received anonymous e-mails that upbraided him for "even being associated with that monster Heath!" Despite this, he was not put off by a journalist. "I would like to help," he said. "The problem simply is that many of the people who were actually involved in the work are dead," and he rattled off a series of names.

"There is one person I've never met but who was involved in the early years—Frank Ervin is his name. He left Tulane for Harvard and created a brilliant career for himself, even though, like Heath, he earned a reputation for controversial research. But I don't know whether he is still alive."

It turned out that Frank Ervin *was* alive, but—in his own words—he was hanging on by his fingernails. He had gone to the little Caribbean island of St. Kitts to die. "But I'm still here for the time being. A visit in the next few months would be especially welcome," he wrote me.

I located Ervin, who was then just shy of his ninetieth birthday, through some random pages on the Internet, most of which cursed him as a "notorious experimenter." One belonged to the animal rights organization PETA, which was protesting against the monkey colony he had set up on St. Kitts. PETA claimed that he was profiting from the exportation of innocent animals to laboratories all over the world "for use in experiments in which the animals are tortured and killed." The accompanying pictures showed some small, frail monkeys with glazed eyes and tattoos reminiscent of concentration camps. I could not quite fit this together with a long, storied career in psychiatry, and wondered if they or I had the wrong Frank Ervin. But the PETA folks were adamant that this Ervin was connected to Montreal's McGill University, which listed a professor emeritus in psychiatry trained at Tulane. It had to be him.

A few weeks after our first e-mail exchange, I drove up the dirt road, full of potholes, to the old sugar plantation that Ervin had restored and called his home. A warm, fragrant breeze swept through the main building and made the wooden shutters rattle. I was shown to a living room with a high ceiling. Soon, some groans and heavy, shuffling steps approached the door.

"Whom have we here? Welcome."

Ervin was a large man, with wild, white hair and a complementary beard that gave the impression of a biblical patriarch. But as soon as he smiled, he reminded you of a boy—a sly, wily boy, with a walker.

"Yes," he said, as he slowly lowered himself down on the other side of the long table across from me. "Strictly speaking, I'm only still standing courtesy of the countless pills I take every day, but I refuse to let go. There are still things to do."

Together with his wife, Roberta Palmour, a professor of genetics at McGill, Ervin was overseeing research projects and receiving colleagues who wanted access to his colony's troop of African green monkeys, or vervets. "*Chlorocebus sabaeus* is the new name," he said, explaining how the small monkeys had lived wild on the island since their forefathers arrived from Africa in the 1700s as stowaways on slave ships. The Caribbean climate was favorable and the food plentiful. So, the little population quickly expanded, and the monkeys became pests. The islanders took to hunting and eating them. Many still did.

I sighted some of the sprightly primates with their black faces and olive-green fur in the research colony, which was located a few kilometers from Ervin's home. It resembled what you might imagine an old-fashioned jungle station looks like. Succulent vegetation overwhelmed everything, with stores of bananas and other fruit piling up under the porch roof, and tropical birds gurgling their peculiar sounds. The colony had been set up so that the monkeys roamed outdoors in large airy enclosures, where they lived in natural family groups, rather than individually caged, as was often done at Western universities.

"I believe that anything that has to do with behavior and mental illness must be investigated in a social context," Ervin said, tapping a finger on the tabletop.

His monkeys were particularly renowned in the field of alcohol use and abuse research. Ervin and Palmour had discovered that the animals displayed exactly the same drinking pattern as people when they were given access to booze. Some became decidedly abstinent, while a slightly larger group consumed in moderation. Finally, there was a small group that took to the bottle and got consistently plastered. Ervin said these hard-core drinkers would drink themselves to death if the booze were not taken away from them. The monkeys provided a great opportunity to explore the biology of substance abuse and the damage such abuse does over time.

But Ervin and Palmour were convinced their little vervets

possessed greater promise than this. They were in the process of investigating whether the monkeys could be good animals for research into dementia, for which no good animal model had yet been found. The colony housed groups of monkeys who had lived almost a quarter century, the equivalent of ninety years for a human being. And it turned out that many of the vervets showed some of the same changes in their brains as patients with Alzheimer's—the characteristic protein plaques, which have not yet been found in either rodents or dogs. The monkeys on St. Kitts were about to have various aspects of their mental condition tested to see if they had clinical signs of dementia. "And *then* we are in business," Ervin said in a way that made me think that he had forgotten he presumably wouldn't be around to see the business take off.

It was clear that this colony of just under a thousand monkeys was now the man's life's work. But it was not because he had planned to spend most of his career in the company of animals. That was the fault, he said, of "the 1970s." Psychiatry was under fire, and it became very difficult to conduct research on people. "That trend has continued ever since, and today it is actually so difficult to get permission to do human research that the patients ultimately suffer for it. Psychiatry has not advanced its understanding of disease much beyond what it had forty years ago. I keep up with the literature, and again and again I see completely new studies that look like something we did decades ago—the young people have just not read about it." He felt he had been forced then to abandon the patients who had been the entire point of his research.

I asked about Heath. Abruptly, Ervin twisted his head toward the window and, for quite a long time, stared out at a solitary horse grazing in an enclosure in front of the house. Then, having collected himself, he returned his attention to me.

"It's been a long time since I heard anyone talk about Bob Heath. Of course, there are a lot of people from his generation who never get mentioned, but Heath ought to be well known today. He was one of the first biological psychiatrists. Already at the end of the 1940s, before anybody else was doing it, he said that schizophrenia was a brain disease with a genetic foundation."

Still, Heath was anything but a one-trick pony, Ervin emphasized. He rolled up to Tulane with experimental psychologists, radical young psychoanalysts, and a cultural anthropologist. Unheard of at the time.

"Bob wanted to understand *the person*, and he stood for a holistic approach to the psyche, which can best be described as 'biopsychosocial'—even though that is a god-awful word. But the approach was incredibly attractive to me and others."

So was the man, apparently. As Ervin remembered Heath, the words "charismatic," "elegant," and "aristocratic" came one after the other in quick succession. Heath had been a dazzling teacher who, year after year, won the prize as best lecturer and who, in a just a handful of years, increased recruiting for Tulane's psychiatry department from 0 to 30 percent. "At Tulane, psychiatry was quite simply a respected field, which it decidedly was not in other places at that point," said Ervin.

But it was not just the university that felt the force of Heath. He set out to revolutionize the state of Louisiana's backward psychiatric hospital system. He started psychiatric clinics in the major cities, which was entirely new, and sent his interns out to old, neglected facilities in rural areas.

"Bob helped raise the money for the new state hospital in Mandeville outside New Orleans, and he insisted on introducing the newest trends in psychiatric treatment. We worked with British and Dutch methods and experimented intensively with talk and other forms of therapy as well as phasing people through halfway houses before they went home. That was at a time before there were drugs, when hospitals were otherwise straight-out storage facilities for the insane. Completely indescribable."

Nonetheless, Ervin described the worst example: the infamous East Louisiana State Hospital in Jackson, where "storage" was in a category of its own. There were the "unclean" wards, one for men and one for women. Each ward consisted of a single circular room with a ring of beds covered by rubber sheets and windows so high up no one could see out of them. The patients were so far gone that they could not control their bodily functions or figure out how to use a toilet. So the shit and the piss flowed freely. Once a day, a man came in and hosed down everything—the room *and* the patients—with cold water.

Among the young psychiatrists, there were stories about the "dungeon" in the basement of this dilapidated building, halfway underground. This was where they kept 120 of the most unmanageable patients, their arms and legs in chains, each in their little hole covered by a grate through which their food bowls could be lowered.

"It sounds like a parody of the Middle Ages, but it was a reality in America in the year 1952."

Heath made them send some of the long-term, chronic patients to his ward at Charity Hospital, where he attempted to do something for them—some kind of social rehabilitation in the hope that they might be able to be released into society.

"We tried. It didn't have much of an effect, but it shows his humane approach," said Ervin, nodding slowly. "Bob was a new broom that swept out the old cobwebs, and if you just look at his achievements as a clinician—well, he was a hero."

Between 1954 and 1958, Ervin was an intern in psychiatry at Tulane. He was quickly tapped to be a member of Heath's research team and, later, his protégé. "Bob's golden boy," as he called himself. This meant that the master might call him from the lab in the middle of the night to discuss his most recent observations and theories, and that Ervin, who at that time had four small children at home, would rush off early the next morning to work on the case.

"He was absolutely brilliant, make no mistake. And his knowledge was colossal. The man simply read *everything* and was up-to-date with the latest news, whether it was in anatomy, physiology, or biochemistry. And he initiated the first at least halfway thought-through attempt to do something for the enormous population of desperately ill people who were otherwise just given up on."

At this I decided that I had to raise the critiques I had seen of the experiments, that they were haphazard, and that the controls left much to be desired. Ervin frowned at me.

"You have to understand that the research context in which Heath was working was very different from today. He was actually responsible for some excellent initiatives—he systematically tested low and high currency strengths, different pulse values, and that sort of thing. In other words, he did the obvious controls you could do then. But there were other things you could not control for. He couldn't take a bunch of healthy people and stick electrodes in them, could he?"

Ervin turned in his creaky chair and looked me square in the face. He wanted me to understand one thing: A lot of animal experiments were done before they ever started work with human patients. When they finally felt confident trying a method on patients, they observed plenty of interesting effects. The septum, on which Heath focused, was an area with good access to and connections with the amygdala, where large portions of the emotional brain were located. At the same time, the septum had a direct connection to the hypothalamus.

"Stimulate there and you fire up a number of emotional brain areas, and at the same time get stress hormones and all sorts of juices going. The research program was certainly not blind. Bob saw an abundance of effects that convinced him that he was on the right path. He even got a rise out of some of the most withdrawn schizophrenics. Imagine it—for them, it might be the first time in thirty years they had shown any kind of affect! Sometimes, it might be anger that came to the surface, but at least it was *emotion*."

At once, a broad smile crossed Ervin's face as he was reminded of a patient from those days. The man, a longtime patient with a diagnosis of paranoid schizophrenia, had been inducted into the electrode group, and he improved steadily with treatment. Gradually, he became "somewhat functioning," so that the doctors allowed him to leave the hospital from time to time. "The first thing he wanted was a proper haircut. He waltzed into a little local parlor, took the white stocking cap off his head, and asked the barber to please cut around the electrodes."

For a moment, this absurd image hung in the room between us. Then, Ervin grew serious.

"I think Heath's tragedy was that he suffered from magic-bullet syndrome. He wanted to find *the* solution, and it made him a little too uncritical of his own work. But I don't entirely know what happened when it all fell apart for him. That was after my time. But maybe you can find out from Charles O'Brien. He was a student and took over my position as Heath's favorite in those years when things started to go south. Chuck is a fine fellow."

I could well believe it. At the very least, O'Brien's reputation and his CV were impressive. For many years, he had been the head of psychiatry at the University of Pennsylvania. He was in his own right considered one of the leading experts on addiction and substance abuse. So, I hopped on a plane from St. Kitts to Philadelphia to meet with him. We did so at the university center bearing his name, the Charles O'Brien Center for Addiction Treatment.

"How's Frank's health?" O'Brien began, after a firm handshake. At my report, he shook his head. He was approaching eighty himself. With his dark hair and slender shape, he did not look it. But when he moved, it was with stiff legs and a slightly curved back. "I was just in Europe with my grandchildren . . . that's the last time I'll travel so far."

O'Brien had been an intern in Heath's lab at the beginning of the 1960s. He readily admits Heath's influence in his career. Like his mentor, he had trained as both a neurologist and a psychiatrist, and believed the two specialties were naturally and inextricably intertwined. He also credited Heath as the source of the foundational thinking—that chemical dependence was a physiologically driven behavior rather than a personal flaw arising from social factors or a bad childhood.

"I have no better way of saying it than that Bob had some really good ideas, and he was far ahead of his time. But things were incredibly rigid in those days. I remember how, well into the 1960s, I would get myself into heated discussions with people who were completely incapable of conceiving that it was the *brain* that was disturbed in people with schizophrenia. No, it all had to do with 'something in the environment.'"

O'Brien's office was a generous space with high ceilings and the feel of a library reading room. We had been seated on comfortable sofas in a corner walled off by filing cabinets and stacks of journals when suddenly O'Brien stood up. He went to the back of the room where bookshelves rose almost two stories high and began to climb a spiral staircase to a balcony midway up. Continuing to talk with me, he moved slowly along the shelves up there, clearly looking for something. After going through a few yards of book spines he looked down and asked whether I had managed to get hold of the films.

The films?

I had written to Tulane in the hope of seeing some old papers or, perhaps, a yellowing patient journal. I had no idea that the institution possessed an archive of 16mm films, and later videos, shot by Heath during his experiments dating from 1950 and stretching into the 1980s. A treasure of celluloid strips. He had been scrupulous in documenting each patient interview, surgery, and reaction. He sat and talked with his patients while they were being stimulated, while simultaneously capturing the activity taking place deep in their brains.

"Very innovative for the time," said O'Brien.

"In the meanwhile, you should read this," he said from close to the ceiling, where he had finally found what he had been looking for. He slowly climbed down and handed me a book with pigeon-blue binding. On the spine, it read: *The Role of Pleasure in Behavior.* It was a sort of proceedings of a symposium held at Tulane in 1962. A small group of researchers—including a number of then internationally renowned names—had convened to consider how aspects of pleasure regulated human behavior. I flipped through the book. They had shared the results of animal research, and experiments with patients, and even invited the perspective of a philosopher.

"It comes out here that treatment was the rationale for the operations and the many experiments, but . . ." said O'Brien, shrugging ever so slightly, "Bob's interest in the role of pleasure sometimes took other forms that are difficult for some people to handle."

Unfortunately, he could not give me a good answer as to why

his mentor, who embodied such great ambitions and invented utterly new methods, had been so thoroughly ignored. He had been at the lab after Frank Ervin, but he had moved to England early in his career and lost touch with the group at Tulane for a long time. On the other hand, there was one student—an Arnold Mandell—who had been very close to Heath, almost obsessed with him, he recalled. Mandell had worked through his experiences at the lab in a sort of roman à clef, which supposedly described what had gone wrong for Heath, including what had led to his ignominious fall.

My ears pricked up. O'Brien did not know where this Mandell lived today, but he believed it was certainly worth my while to hunt him down.

"In the meantime, I'll see whether I can track down his manuscript. Who knows?"

CHAPTER 3

Treat Yourself

Roy was standing at the edge of the roof and his psychiatrist was a few steps behind him pleading with to him to step back. Robert Heath had run up the stairs and was now standing, breathless, with his lab coat fluttering on the roof of the university building. He spoke gently and patiently, extending his hand toward the young man, who was teetering on the roof's edge above Canal Street. The man turned slowly away from the psychiatrist's outstretched arm and looked demonstratively down into the street below. He leaned out precariously and threw a quick glance over his shoulder toward his doctor.

It was not the first time there had been a commotion around Patient B-10. Some of the younger interns whispered about how the sullen, moody patient deftly manipulated his doctors. The last time he apparently grew weary of life, the ward got a call from a couple of cops. They had taken the man into custody on a bridge over Lake Pontchartrain, where he had been threatening to drown himself in the shallow water.

Charlie Fontana, Heath's trusted EEG specialist, drove out at once to bring the patient back. Fontana had been working with Roy for a long time. He had welcomed him into his home on several occasions and knew him well. On the car ride back, Fontana turned around in the front seat and asked Roy whether he had seriously meant to jump in the lake.

"Nah, not really," replied the patient, without blinking an eye, "my electrodes would get rusty."

The electrodes in his brain were quite precious to him. The story went around the ward about the time Patient B-10 tried to sell himself as a guinea pig. One morning, he took off from New Orleans and made his way to the University of Chicago Medical School and the well-known psychiatrist Daniel Freedman. Patient B-10 introduced himself, took off his little white cap, and bent forward so the rather bewildered Freedman could observe the electrical gadgets in his scalp. He then asked whether the doctor might not be interested in getting access to them. For $5,000 Roy would offer his services as a research subject.

Patient B-10 was part of a new generation of electrode patients at Tulane. The twenty-five-year-old was not a classic schizophrenic, like many of his predecessors, but an epileptic. For years, he suffered from psychomotor epilepsy, as they call it, which comes with a number of symptoms that originate in and spread out from the temporal lobes—not just regular seizures but also recurrent and "brief episodes of impulsive behavior" during which his conduct could sometimes approach the psychotic. At this point in the late 1950s, there were no drugs to keep this in check. So, Roy's neurologist sent him over to Robert Heath in the hope that the professor's unorthodox methods might have some luck.

The Tulane group had come a long way since they first leapt out of the shadows at the now notorious schizophrenia symposium back in 1952. Not because the conference was a success—to the contrary. Not only Robert Heath himself but the entire group was surprised at how much resistance they encountered from older, established colleagues about their way of thinking and the experiments they conducted. Yet it did not shake Heath's faith in the larger project—not in the least. No, it galvanized him. He *knew* he was onto something, and he just had to get his message out.

At first, it was only the local newspaper, the *Times-Picayune*, which was mightily impressed by the university and sent its journalists out to do big stories on the psychiatric revolution that seemed to be taking place at Tulane. But soon, important national media were also brought on board. In 1953, *TIME* came to visit and did a profile of the dapper Dr. Heath. In the magazine—which Heath kept on display in his office for many years, he was called "the Gregory Peck of psychiatry," a pioneer bringing the hope of a cure to the mentally ill.

Two years later in 1955, a crew from the popular CBS television series *The Searchers*, which highlighted select, cutting-edge research, came down from New York to do two whole episodes on Heath. These programs made up the big season finale, and their footage of patients being stimulated by electrodes planted directly in their brains created an immediate sensation. No one had ever seen anything like it before. They also showed Heath arguing passionately that biological psychiatry was about to solve the puzzle of how and why we human beings break down from psychological illnesses. Illnesses that "are still shrouded by stigma and guilt, reactions that are always borne by ignorance," he said with his gaze fixed steadily on the camera lens.

Thanks to this exposure and his high profile in New Orleans, Heath no longer needed to go out and hunt for patients. They came to him—either on their own or through their families and doctors. Heath kept a file folder in his office, filled with letters from disturbed people convinced that what they needed was some electricity shot through their brains, and he received referrals from colleagues all over Louisiana. Sometimes, it was a question of choosing among eager volunteers.

The psychiatric and neurology ward had become a well-tuned, well-oiled machine. They had equipment other hospitals could only dream of, thanks to a torrent of research funding. There were public funds, but most of the money by far came from private institutions and wealthy patrons in New Orleans. Robert Heath's combination of charisma and expertise made him an extremely effective fund-raiser. He got not only foundations but also local fat cats to reach deep into their pockets. It was no coincidence that Heath quickly became a highly sought-after clinician who treated an array of the city's bigwigs and their nearest and dearest in his private practice. He was an empathetic but demanding therapist and this worked with both manic industrial magnates as well as their maladjusted daughters and depressive wives. In time, even celebrities found their way to his couch, figures from American history. There was the manic-depressive district attorney of New Orleans, Jim Garrison, who became notorious for arresting local businessman Clay Shaw for the murder of President John F. Kennedy (portrayed decades later in Oliver Stone's film *JFK*). Robert Heath was also the psychiatrist they asked to attend to and evaluate Louisiana's controversial governor Earl Long, when he was committed to the state hospital at Mandeville in the 1950s. Heath diagnosed a manic disorder—possibly triggered by amphetamines. Long reacted by dismissing the entire hospital board and then absconding from New Orleans with the stripper Blaze Star.

Back home, Robert Heath acquired a peach of a partner: Irene Dempesy, the secretary who would stick with him through thick and thin until retirement several decades later. She came from a career as a stewardess. But at Tulane, she nestled right into the center of the administrative web and kept jealous guard of all the threads. Irene was surprising in every way. She was beautiful and elegant, with well-coiffed dark hair, long legs, and a figure that was dangerously close to the hourglass ideal of the day. And she was incredibly efficient and in the know about everything that was going on.

"Irene is Bob's thalamus," his staff murmured, referring to that region of the brain through which all information passes before being processed elsewhere. No one said it out loud, but on occasion she also acted as his frontal lobes with their cool, thoughtful superego. If Bob unleashed his temperament and screamed at a student that he was "a total idiot completely devoid of any creativity" or blew his stack and threatened an employee with a pink slip, it was Irene who smoothed things over with the stunned victim.

"He doesn't mean it. In fact, Bob needs to be contradicted. He just needs time to think it over."

The two became an inseparable unit: Bob-and-Irene. They spent hours in each other's company every day. As they strode the halls, side by side, they looked like a royal couple on a tour of their kingdom.

It was a realm of oversize personalities, innovative research, and copious funding, and it attracted gifted young people with ambition. Tulane's psychiatric ward was only ten years old, but no other university in the country lured so many medical students to the field. Almost 1 in 4 each class year wanted to specialize in psychiatry, and year after year, Robert Heath was selected hands down as the best teacher at the university. Except once, when the honor went to his psychiatrist colleague Ruth Patterson, which infuriated him and made him exert himself even more.

Heath was able to speak the language of psychoanalysis and guide students through its airy concepts and interpretations—only to turn from the chalkboard and assert that, while it was all very interesting, "it just won't cure patients."

Elsewhere, the field of psychiatry was considered almost esoteric, not really something for most diligent medical students. Here, the discipline was far more tangible. It had to do with brains and grappled with the nervous system itself. It was anatomy and physiology and, therefore, entirely within the model of classical medicine.

At the same time, something happened that changed the game for a strictly medical approach, something that would eventually allow it to reclaim psychiatry from the psychoanalysts. The first antipsychotic medication came on the market, and over the next decade it helped empty psychiatric hospitals throughout the Western world. In 1950, while Robert Heath was implanting the first electrodes into his schizophrenic patients, the French pharmaceutical company Rhône-Poulenc produced the drug RP4560 (Chlorpromazine). In small, preliminary experiments at state hospitals in Canada, the compound proved to be sensationally effective at bringing even the most raving patients out of their acute psychoses. In 1955, it was marketed under the name Thorazine.

In the United States, Heath and his people were among the first to test the new wonder drug systematically. The National Institute of Mental Health provided generous funding for that purpose, and suddenly, from one day to the next, a whole room at Tulane was filled with boxes of the psychoactive drug.

In particular, the pills were used at the old East Louisiana State Hospital in Jackson. "A horrible place, a pure storage facility," admitted Donald Gallant, the newly qualified psychiatrist who headed up the medical experiments. Its dilapidated, prisonlike buildings housed more than five thousand patients. These wretched inmates remained there for an average of twenty years. Many ended their lives there. It was the chronically ill—typically, schizophrenics for whom nothing more could be done—who wound up at this facility in Jackson. People whose families often forgot all about them because insanity was a taboo subject, treated like a dark secret.

But in the middle of this hellish "loony bin," Robert Heath established a special ward with 130 beds, where Thorazine testing would take place. And month after month, an ever-increasing number of patients who had been strapped down, bedridden, and given up on long ago, were able to get up and walk. Gallant began to think of the new drug as a sort of Thor's hammer that shattered psychoses. And in a few years, 2 out of 3 patients were well enough to leave the hospital.

Thorazine and its striking effects supported Robert Heath's belief that schizophrenia really *was* a biological ailment and an illness that resided in the brain. But might the drug also call into question the use of deep brain electrodes?

There was no rationale for cutting and poking around in patients' brains when all you have to do is give them a pill. After treating around thirty-seven schizophrenic patients, Heath was forced to recognize that this group did not get much out of electrode stimulation. The very withdrawn patients in particular could be very frustrating to deal with. Heath and his staff could *see* in them that something changed when the electricity was turned on—and that it did the patients good. They woke up and were, generally speaking, more normal. But when you asked them, they never *said* that they experienced any difference. As he repeated, "It is as if they lack a language for pleasure."

However, this did not mean the abandonment of the theory that there was something wrong in the patient's septum. To the contrary. This area of the brain proved, indeed, to behave abnormally in schizophrenics. Measurements from deep electrodes revealed again and again that the septum went amok in frenzied activity when patients were acutely psychotic and hallucinating. This activity was not found in the control subjects researchers studied—cancer patients, arthritis patients, and a few with Parkinson's disease—so it seemed to be specific. Heath was certain that the septum had to play a central role. And because a drug like Thorazine remedied some of the psychotic symptoms for extended periods of time, he began to think in the direction of biochemistry.

Heath hired a biochemist and started collecting blood from patients and injecting monkeys with it. The animals appeared to react with schizophrenic symptoms. He saw it as a decisive discovery. He didn't know that these would be the seeds of his scientific and academic destruction.

He continued his work with electrodes, but the project took a turn along the way or, rather, expanded. His patients were different. They were still people for whom traditional treatment did not work, but they were epileptics with serious behavior problems and severe depression. The primary purpose and the reason they even came to Tulane, Heath always said, was for *treatment*. There was a therapeutic aim. But from the scientific papers that flowed in a steady stream from the group, it appeared that the patients were also test subjects in a far larger project.

Heath's ambition was not exactly modest. He wanted to solve one of mankind's most basic mysteries—to understand the relationship between *mind* and *brain*. In practice, this meant that he had to try to characterize the function of different areas of the brain and figure out how they worked together.

For this, the electronics were a godsend. The researcher could sit across from someone who was feeling something or thinking something or doing something and track, via EEG, what this did to his or her brain activity. They could also stimulate the brain in one place and see what happened to the activity in other areas over time. Moreover, they could ask their guinea pigs to describe their own subjective experiences.

For Robert Heath, it was about pinning down emotions. He was almost indifferent to thoughts—which, with all their elegant rationality and hard logic, belonged to the highly developed cerebral cortex. But primitive emotions were an uncharted country. Emotions could overwhelm even the most sophisticated individual. As he noted, "The most accomplished scientist cannot ponder the intricacies of a scientific equation when he is in a highly charged sexual state or in a state of rageful dyscontrol or panic. In other words, he, too, is sometimes influenced by primitive, self-centered, often unreasoning, emotional thought that cannot always be sublimated by plunging into a dedicated effort."

It was a matter of mapping and exploring the deeper parts of the brain and finding out what they do. Just a few years before, this sort of thing would have been a closed book. Only the very few even imagined that individual feelings might have their own seat in the brain, much less one you could find experimentally. Certainly, people had poked into rat brains and discovered motor skills and other sensory regions, but human emotion seemed different, something far more complex and intangible. A vague and difficult phenomenon to deal with.

However, Donald Gallant had been working with patients for years—talking them through, like a good psychotherapist, old and musty but distinctly emotional recollections while the EEG machine was running at full throttle. After a stint as a psychiatrist for the US Air Force, Gallant had just returned to Tulane, and he volunteered to assist with the electrode patients. He could *see* with his own eyes how the same brain structures were set in motion when different people relived their completely different pasts.

The day before they met, patient B-10—Roy—had watched a television program about juvenile delinquency, which made him tell Gallant about some horrific experiences from his childhood. Egregious violations and even more egregious punishments. These recollections showed up in his rostral hippocampus and his amygdala as high-frequency peaks of activity until Gallant gave him a math problem to solve. The logic demanded by the numbers engaged his thoughts and calmed him down at once, but the previous pattern of activity returned as soon as the psychiatrist took him back to his childhood.

The same patterns recurred with all sorts of recollections—regardless of content, no matter whether they were positive or negative. But what really surprised both Gallant and his boss was that the recollections always belonged to a *distant* past. Episodes dealing with a patient's present and current situation did not appear in the same regions of the brain. And it was a pattern that also appeared when the situation was reversed. When, without knowing it, B-10 was stimulated by the electrode in his rostral hippocampus, memories suddenly rushed out but always about situations that happened years before.

Don Gallant felt privileged. A young and untried researcher, he had free access to secret chambers, doors no one had ever opened before. To anyone who cared to hear about it, he described his experiences in the little interview room on the second floor as "utterly wild."

It was just as wild for another of Heath's trusted employees, psychiatrist Frank Ervin, who experienced the "aha" moment of his life in Tulane's operating theater. One day, out of the blue, a patient lying on the operating table suddenly struck out with his arm and overturned the surgeon's sterile instruments. Ervin had stimulated an area in the brain called the tegmentum, and the man reacted with instant fury.

"What's happened?" Ervin asked, but the patient himself was completely surprised and could not explain it. He apologized but, for some reason, could not control himself. Ervin stimulated him again; once again, the man on the table felt an explosive rage that simply *had* to find an outlet.

This was something entirely new. Ervin and Heath only intended to test on people what colleagues had found in animal research. The famed Horace Magoun had shown that the brain's so-called reticular formation played a role in the level of consciousness—that is, how alert and attentive one is. Ervin had expected that, with a little stimulation, they might be able to raise their patient's alertness level, to get him up on his toes, so to speak. But this—uncontrollable aggression on command—this was unexpected. It socked Ervin right in the gut: If you could map the internal circuitry of aggression, you might be able to find a way to curb it.

They had another patient, A-10—Joe was his name—whom they asked to recall something unpleasant, a situation that made him furious, while they measured the data from his fourteen electrodes. His recollection provided a characteristic signal from the electrode in his hippocampus—violent oscillations of quick activity that are called spindles. The next step was to send a jolt of electricity through the same electrode. And then something happened. Joe—who otherwise possessed a calm, pleasant demeanor—suddenly distorted his face into a grotesque grimace. One eye rolled back into his head, while his body writhed as though something were giving him great pain.

"It's knocking me out . . . I just want to claw . . . I'll kill you . . . I'll kill you, Doctor," he said. Then, it was over. In the back, Charlie Fontana cut off the electricity and, just as abruptly as the bizarre attack began, it was gone. The intense anger and lust to kill was the direct consequence of 2.5 milliamperes of stimulation to the lateral part of the hippocampus. "But why did you shout, what were you feeling, and why were you angry at the doctor here?" the shocked psychiatrists began to ask. But the man himself had no idea.

"I don't know why I said that. I don't have anything against him, he was just there."

A-10 was always mentioned in Robert Heath's public speeches when he wanted to show the listeners how close pleasant and painful emotions are to each other—and how they can be produced on command. His descriptions made many people uneasy. In fact, they made some of his colleagues wonder whether the research at Tulane was even defensible, much less desirable. Not only did Heath and his people put an innocent patient into an excruciating situation, but the whole strange setup touched on something fundamental. Quite literally, Heath and his people were demonstrating that you could control and shape a person's feelings and actions.

The discovery shaped Frank Ervin's career for many years to come. Having been Robert Heath's favorite and a protégé for a number of years, the tall, sarcastic Texan was recruited by Harvard University, which wanted his expertise with electrodes. Ervin joined up with a neurosurgeon named Vernon Mark. At several hospitals in Boston, the two began working with a very special group of patients—people who frequently exploded into violence.

In the meantime, the research Ervin left behind at Tulane turned more and more toward pleasure, desire. *Hedonia*, the internal primordial force, continued to haunt Robert Heath. He wanted to penetrate to its essence, to map its physiological mechanisms, and to understand exactly how it influences the psyche and the way people live their lives. But there was also another side to the research: If you could harness the power of pleasure, you could use it to modify unwanted behavior.

Nobody who knew him was in doubt that Bob himself was a bon viveur. Here was a man with an unusually high energy level and a huge appetite for everything good in life.

"He's the kind of hypomanic who simply doesn't know how depression feels," said Frank Ervin. And it was clear to everyone that the boss got high on his research—he showed up at work before the devil could get his shoes on, eager to get started and bubbling with ideas until he got home at night.

"Work Hard, Play Hard" might be a slogan invented for Robert Heath. He played tennis and golf with gusto. He frequented the best restaurants in town, drove expensive cars, and loved parties. He even called his summer house (a small farm out in the country near Picayune, Mississippi) Hedonia. The whole family—now with five children—spent its weekends there. Heath hunted and bred cattle, and otherwise sat back contemplating the cow patties with a straw between his teeth. Once or twice a year, he gathered his entire psychiatric staff with wives, children, and significant others for "a day at Hedonia" with an atmosphere of abandon, volleyball in the grass, barbecue, and plenty to drink.

While hedonism was admired as a lifestyle in warm, sociable Louisiana, the scientific study of pleasure itself was a bit harder to accept. And since Robert Heath believed (entirely in line with his psychoanalytic mentor, Sandor Rado) that sex might very well be an excellent treatment for a patient, he was on a collision course with the outside world. The atheist Heath despised the Puritanical prudishness he encountered among the Catholic sisters who still ran Charity Hospital; the nuns, in turn, were not happy with the way he practiced psychiatry. They did not care for his electrode experiments, which they believed bordered on sacrilege. Taking control of a human soul with machinery?

But according to the agreement between Tulane and Charity, Professor Heath had a monopoly on treatment. So the sisters couldn't take the electrodes out of his patients' heads. What they could do—and did many times—was rip patients from the arms of their spouses. Whereas Robert Heath insisted that patients should be allowed to use their private rooms to go to bed with their better halves if they wanted, the nuns could not tolerate that sort of thing. It didn't belong there, it was sinful, and it was against the rules!

More than once, Robert Heath was summoned by loyal nurses or young doctors dealing with hysterical or shocked patients who had been caught in flagrante by the sisters or their spies. In one such case, he made the long drive in from Hedonia on a late Sunday evening to take care of a schizophrenic woman. She had finally begun to come out of her psychotic withdrawal, but now the nuns' intervention had knocked her back to square one. On the way into town in his big Buick, Heath riled himself up more and more for the confrontation he planned with the ossified old prioress.

At the ward, he rushed to his patient. Trembling, silent, wrapped in a sheet, she sat crumpled in a corner and had to be told again and again that she didn't do anything wrong. To want love—even physical love—was only natural, he explained calmly and insistently. Pleasure was good, and it was healing.

But the more Robert Heath dug into its mechanisms, the more surprising and mysterious they seemed. In 1962, he gathered a small circle of interested colleagues to a symposium entitled *The Role of Pleasure in Behavior*, and he himself presented the most noteworthy findings. Once again, he had gone a step beyond anyone else. Recognizing that the brain functioned not only through electrical signals but also through chemical communication, he began to apply drugs directly into his patients' quivering brain tissue. While other researchers were satisfied with medicating their subjects by way of pills or injections, Heath wanted to go straight to where brain cells signal one another.

Only a little was known about the molecules cells exchange and how they do it; but, from regular medical practice and animal research, certain drugs were known to have an effect on the brain. At Tulane, they decided to test a total of fifteen chemical compounds that covered everything from sleeping pills to morphine to histamine to adrenaline and a couple of drugs identified as neurotransmitters. Precisely measured doses were applied to the septum, hippocampus, or thalamus through hair-thin glass tubes that were implanted as if they were electrodes. The test subjects were ten patients with severe psychomotor epilepsy, which unleashed not only cramps but also psychosis-like symptoms. As Heath told his colleagues at the meetings, he had previously discovered that electrode stimulation in the septum might curb or even eliminate attacks. But the effect was only temporary—it might last weeks or months. Therefore, he presumed that there must be something wrong with the chemistry in the affected area of the brain and that the way to test this was to apply some chemicals directly and see what happened.

And things were happening with acetylcholine. It was one of the brain's own neurotransmitters and the only drug of those tested that provided a pronounced experience of pleasure—in fact, often a pleasure that had a clear sexual tone. In one patient, Robert Heath told his listeners, it turned into actual orgasms.

He showed EEG readouts and film recordings of B-5, a woman, thirty-four years of age, who suffered from a severe form of epilepsy for twelve years. She experienced both grand mal attacks with cramps and unconsciousness and psychomotor episodes in which she screamed and shouted, pounded her head into the nearest wall, and lost orientation. Even though the doctors pumped her full of every available medication, she still had eight to ten attacks a week and was desperate for a treatment that worked.

At Tulane, they had worked for some time with animals and put glass tubes into cats and monkeys. Now, they did the same to B-5. Along with the usual electrodes that could measure what happened deep inside, Heath and his surgeon lowered the thin glass needles into the tissue and injected precisely measured doses of acetylcholine into the patient's septum. A single dose—just five micrograms—unleashed a marvelous chain of mental reactions that almost unfolded as a fixed behavioral program. This happened ten out of the twelve times they did the experiment, and it began with the woman, in the first few minutes after the injection, experiencing a lift in her mood. She clearly showed more interest in her surroundings and chatted away with everyone in the room. After fifteen minutes, she found herself in a state of "mild euphoria" in which she could more easily than usual solve various math problems the researchers gave her. At the same time, she introduced sexual motifs into the conversation. After another ten to fifteen minutes, the session culminated in repeated spontaneous orgasms—a phase that could last up to ten minutes.

Robert Heath described in clinical terms how this singular treatment took place once a week for twelve weeks and that, in that time, B-5 did not experience a single epileptic fit. In fact, they stayed away for six months. Then, she suddenly had several attacks, one after the other, but something else happened. The disease could now be controlled by a medication that had had no effect *before* the woman received doses of acetylcholine. At the same time, Heath remarked in conclusion: "This patient, who is now married for the third time, had never experienced orgasm before she received chemical stimulation of the brain. But, since then, she has consistently achieved climax during sexual intercourse."

Jaws dropped among the visiting colleagues, and Robert Heath admitted flat out that he had no good explanation for what happened. One could understand the immediate effect of a drug on the brain, but it was not logical in and of itself that it should work for months at a time. He also realized perfectly well that it was not practical to use direct injections through glass needles as a standard treatment. Of course this was a purely experimental approach that could only be justified in a few serious cases. But, he argued, the exploration of the phenomenon of pleasure itself might yield results. It could accelerate the search for medications that might make people feel pleasure and increase their level of attention without a complicating factor with which they were already familiar from the prevalent drugs of the day—namely, addiction.

Generally speaking, the ability to provide people with pleasure at the right time was the path toward treating abnormal psychology. Every sort of antisocial or neurotic behavior, as Robert Heath saw it, was the result of an early misprogramming of what was felt to be good or unpleasant, respectively. Psychotherapy poked around in how this misprogramming might have happened—childhood circumstances, for instance—but did not work very well as treatment. What was needed was reprogramming. Theoretically, one could imagine creating momentary pleasure or joy in the patient when the person, for example, felt anxiety or rage without having any reason for it and thereby eradicate the undesired feeling.

This time, just like back in 1953, a book came out of Robert Heath's symposium; but unlike the first time, it received positive reviews. In the journal *Psychosomatic Medicine*, a reviewer especially praised the Tulane group's audacity. As he wrote, "The direction of this research over the intervening 10 years from a primary concern with a disease entity to a primary concern with a basic human experience reverses the usual trend and represents a welcome contrast in a period in which applied research threatens to crowd basic research out of the picture."

Nevertheless, there was something about pleasure that plagued Robert Heath. Namely, that *he* was not the person known all over the world as the man who discovered and identified the pleasure center of the brain. The feather in that cap went to the Canadians James Olds and Peter Milner from Montreal's McGill University because of a groundbreaking 1954 article they published on ecstatic rats.

Olds and Milner were hunting down the places in the brain that provided rewards, and they did so by putting electrodes in different areas of the brain in sixteen lab rats. The rats were put into a so-called Skinner box, where they could turn on their own electrodes by stepping on a little pedal. In six of the animals, the researchers felt they had really stumbled onto something. These rats eagerly stepped on the pedal up to 80 percent of the time. For a single animal, it was up to 92 percent. When Olds and Milner eventually cut off the rat's head and dissected its brain, they discovered that the tip of the electrode had ended up in the ventral thalamus.

Eureka! The pleasure center was found, and its discovery reverberated throughout academia.

In a way, Robert Heath had come first with his schizophrenic patients but, in his descriptions in 1952, he had not focused on pleasure. First of all, he was on another mission—it was about arguing for a treatment of schizophrenia—and, second, the reaction of the patients to the stimulation was not as pronounced as in the rats. Now, he could see that schizophrenics had the least reaction to stimulation while other patient groups reacted with far clearer pleasure.

He harbored a deep, dark frustration at being beaten to the finish line but only shared it with a few trusted colleagues. Outwardly, he went about his business unfazed. At any rate, he had something the others did not: His research animals could talk. They could give him access to something quite crucial, which was the inner, subjective experience. If one were ever to hope to link specific parts of the physical brain to the airiness of the psyche, there was only one way, and that was to go to the living, feeling human being. And why not draw inspiration from his colleagues in Canada and their rats? Down in the machine shop, Heath's technicians were asked to construct a little transistor unit, so the patients could stimulate themselves—entirely in accordance with their own desire and need. The contraption had three buttons, each of which was connected to a different area of the brain; each time they pressed a button, a half-second pulse was transmitted.

One of the first to be offered a chance to test the apparatus was patient B-10, and Roy jumped at it. He already had seventeen electrodes scattered about his limbic system and a few areas of the cerebral cortex. And because they each had several live contacts, his brain could be stimulated in a total of fifty-one different positions. One at a time or three in combination. In order to test them all, Roy met with Robert Heath and Charlie Fontana in the laboratory for a six-hour experimental marathon. The subject himself controlled the stimulation and described to the best of his ability the effect each electrode had on him. The only place that felt really unpleasant was the right hippocampus, which made him feel "sick all over." He tested it twice, and that was enough. With the electrode far back on the right side of the septum, the situation was completely different—here, the stimulation was able to eliminate "bad thoughts" and, at the same time, give B-10 a "wonderful feeling" with some pronounced sexual undertones.

Nevertheless, it was the left centromedial thalamus he stimulated the most—just under five hundred times in an hour—even though the stimulation made him very irritated and peeved. When he pushed the button and sent short pulses into his thalamus, it was as if he were always on the verge of grasping a particular recollection, a memory that he felt was important but kept getting away. He pushed and pushed his button but only got frustrated until, after many attempts, he found a sort of solution. By stimulating two places that were associated with reward—namely, the septum and the mesencephalic tegmentum—*while* he pushed the button to the thalamus, he could chase his mirage while curbing his frustration.

When the experiment ended, Roy lit the big cigar he had demanded and gave his psychiatrist a wry, wolfish smile.

"I might want to buy this little box from you, Dr. Heath, and take it home with me."

CHAPTER 4

How Happy Is Too Happy?

It is a good question, but I was a little surprised to see it as the title of a research paper in a medical journal: "How Happy Is Too Happy?"

Yet there it was in a publication from 2012. The article was written by two Germans and an American, and they were grappling with the issue of how we should deal with the possibility of manipulating people's moods and feeling of happiness through brain stimulation. If you have direct access to the reward system and can turn the feeling of euphoria up or down, who decides what the level should be? The doctors or the person whose brain is on the line?

The authors were asking this question because of a patient who wanted to decide the matter for himself: a thirty-three-year-old German man who had been suffering for many years from severe OCD and generalized anxiety syndrome. A few years earlier, the doctors had implanted electrodes in a central part of his reward system—namely, the nucleus accumbens. The stimulation had worked rather well on his symptoms, but now it was time to change the stimulator battery. This demanded a small surgical procedure since the stimulator was nestled under the skin just below the clavicle. The bulge in the shape of a small rounded Zippo lighter with the top off had to be opened. The patient went to the emergency room at a hospital in Tübingen to get everything fixed. There, they called in a neurologist named Matthis Synofzik to set the stimulator in a way that optimized its parameters. The two worked keenly on the task, and Synofzik experimented with settings from 1 to 5 volts. At each setting, he asked the patient to describe his feeling of well-being, his anxiety level, and his feeling of inner tension. The patient replied on a scale from 1 to 10.

The two began with a single volt. Not much happened. The patient's well-being or "happiness level" was down around 2, while his anxiety was up at 8. With a single volt more, the happiness level crawled up to 3, and his anxiety fell to 6. That was better but still nothing to write home about. At 4 volts, on the other hand, the picture was entirely different. The patient now described a feeling of happiness all the way up to the maximum of 10 and a total absence of anxiety.

"It's like being high on drugs," he told Synofzik, and a huge smile suddenly spread across his face, where before there had been a hangdog look. The neurologist turned up the voltage one more notch for the sake of the experiment, but at 5 volts the patient said that the feeling was "fantastic but a bit too much." He had a feeling of ecstasy that was almost out of control, which made his sense of anxiety shoot up to 7.

The two agreed to set the stimulator at 3 volts. This seemed to be an acceptable compromise in which the patient was pretty much at the "normal" level with respect to both happiness and anxiety. At the same time, it was a voltage that would not exhaust the $5,000 battery too quickly. All well and good.

But the next day when the patient was to be discharged, he went to Synofzik and asked whether they might not turn the voltage up anyway before he went home. He felt fine, but he also felt that he needed to be a "little happier" in the weeks to come.

The neurologist refused. He gave the patient a little lecture on why it might not be healthy to walk around in a state of permanent rapture. There were indications that a person should leave room for natural mood swings both ways. The positive events you encounter should be able to be experienced as such. The patient finally gave in and went home in his median state with an agreement to return for regular checkups.

"It is clear that doctors are not obligated to set parameters beyond established therapeutic levels just because the patient wants it," Synofzik and his two colleagues wrote in their article. After all, patients "don't decide how to calibrate a heart pacemaker."

That's true, but there is a difference. Few laymen understand how to regulate a heartbeat, but everyone is an expert on his or her own disposition. Why not allow patients to set their own moods to suit their own circumstances and desires? Like when Robert Heath gave his patients a self-stimulator.

Yeah, well, the three researchers reflected, it may well come to that—sometime in the future, that is—people will demand deep brain stimulation purely as a means for mental improvement.

They stressed that there is nothing necessarily unethical about raising your level of happiness this way. The problem is the lack of evidence that it is *beneficial* to the individual—particularly in light of the considerable cost of the treatment. Even before battery changes, which are needed every three to five years, and regular adjustments, we are talking $20,000 for the system itself and another $50,000 to $100,000 for the operation and hospital procedures.

Today, we have to ask ourselves where a "therapeutic level of happiness" might lie and whether there are risks and disadvantages connected with higher levels.

It seems the unknown young man with accumbens electrodes didn't buy the argument because, after a short time, he stopped coming in for checkups and vanished without a trace. Maybe he found another doctor who was willing to make him happy.

The story made me think of two personal experiences. One was my own, the other my father's. He was diagnosed as manic-depressive or bipolar late in life. After having tried the various medications people prescribed for that sort of thing, he learned to control his mania. But he never hid the fact that he was restraining himself for the sake of others, for his wife and children, who had a hard time dealing with his manic tempo and torrent of words. He himself was in no doubt that the phase that comes just before you lose control and end up in an almost psychotic condition was sublime.

I don't have manic tendencies myself—unfortunately, I usually add. I only inherited the darker side of the spectrum—

namely, depression, and there is not much good to say about that. But the closest I have ever come to an extended feeling of elevated happiness was created by a treatment for depression. This is a well-known phenomenon that experienced psychiatrists can spend a lot of time talking about. This first time you are pulled out of a depression with an effective antidepressant, you may experience it as a catapult to a higher altitude than you are normally used to. For me, it had to do with over a year's worth of little white tablets that were supposed to provide me with more serotonin. Already in the first six weeks, there was a steep upward spike from daily weeping and chronic ennui to an effervescent sense of . . . I can only call it happiness.

The feeling lasted almost six months, and at that time I believed that this hale and hearty inward disposition was my newfound normal condition. It did not seem in any way artificial. And if I am to be honest, this period, even though it was more than fifteen years ago, is still the best thing I can point to in my fifty years of life. My situation has undoubtedly become far better since—I have adjusted to life better; I have written a handful of books for which I have received praise and prizes; I am no longer alone; and I have—objectively speaking—much more to be happy about than at that time. But I have never *felt* so happy and upbeat as at that time.

Would I choose that state again if I could? There's no doubt. Of course I would. The question is: How high a price would I be willing to pay in terms of side effects?

Right now, I take a pill every morning with 100 milligrams of sertraline, which is one of the most well-known SSRI drugs or "happy pills," as they have been unhelpfully labeled. As my psychiatrist says, it is a dosage that does not solve all problems but keeps me from the black hole I was on the brink of just two years ago. It was about time for me to phase them out for the time being, but I noticed something that made me hesitate: a debate playing out in ladies' magazines—for example when *Harper's Bazaar* in the early summer of 2014 had a column titled "The Happy, Sexy, Skinny Pill." This column put a finger directly on the pulse of a new social trend. It had to do with the antidepressant Wellbutrin, which was "on everyone's lips" because, supposedly, it not only inhibited depression but also made women thinner and hornier. Journalist Sari Botton wrote about how she became thinner without changing her diet and how the wonder drug had transformed "this almost-menopausal, almost fifty-year-old woman into a veritable porn star."

Of course, deep down somewhere, I thought Botton was ridiculous, a typical representative of all middle-aged women who dread getting older more than anything and desperately cling to any semblance of youth. Yet even with my indignation, I still checked to see whether this medication was on the market in Denmark and what it was called. Zyban, it turned out, and in Denmark it is mostly prescribed to help people wean themselves off cigarettes. I also checked its side effects to see whether they were worse or more frequent that those with sertraline. They were actually pretty much the same.

Why not just ask my doctor for a prescription and try the stuff? I would be cheating myself if I didn't even try! I have a hard time with enjoyment. It is a recurring theme in my life and something I have meditated on and still come back to. Enjoyment and pleasure do not come easily.

"Why are you always so dissatisfied?" I've heard this again and again since I was a child—whether it was a birthday, Christmas, or just an ordinary day when something went wrong. In fact, it was not exactly dissatisfaction but, rather, the lack of a sense of satisfaction. Even though I got exactly what I had wanted and asked for, it was never as I had imagined it. Like the boy Kay in Hans Christian Andersen's fairy tale about the Snow Queen, who gets splinters from a troll mirror in his eye and, thereafter, sees the imperfect and ugly in everything. The worm in the rose, and the hair in the soup.

I wondered whether it was Robert Heath's focus on pleasure and desire that was my real attraction to his story. It went right to the core of what being a human in the world is all about. The ability to stimulate selected functional circuits in the brain purposefully and precisely raises some fundamental questions for us.

What is happiness? What is a good life?

Hedonia. There is something about this word. It rolls across the tongue like walking on a red carpet and leaves a pleasant sensation behind. Hedonia might well have been the name of the Garden of Eden before the serpent made its malicious offer of wisdom and insight. And more than anything else, hedonism has become the watchword for how we should live. Life is not about being a good, useful person or making any particular contribution or accomplishing something. It is about having as good a time as possible. I have the feeling that Epicurus, if he miraculously appeared in our current industrialized society, would smile and think his philosophy had been broadly implemented.

We in the privileged parts of the world are so rich and satiated that we long ago scaled the bottom rungs of Maslow's famous hierarchy of needs (physiological needs, safety needs, social belonging, esteem, self-actualization) and now have the very top in sight. When "self-actualization" or being all you can be is the ultimate concern and goal, it becomes natural for us to see happiness as an indispensable accessory and to demand it as a separate item—a neat, prepackaged commodity. And to a striking degree, happiness has become equal to fulfilling immediate desires. The ideal is to enjoy life to the last drop, as they say, and there is something almost shameful about not being joyful or *positive*. We expect smiles and effervescent humor from one another—anyone who is negative is an antisocial individual who must be made to fit in. If material, everyday things cannot keep us upbeat, we throw ourselves, conscience-stricken, into meditation, mindfulness, yoga—whatever can lift us up.

In the meantime, the absence of joy and pleasure—anhedonia—has, in its way, become a popular issue in the wake of the disease depression. A quarter of us are affected by it over the course of a lifetime, various studies suggest, and its frequency is increasing in the industrialized world. The treatment of depression has become both a window display and a battleground for deep brain stimulation.

It was with the American neurologist Helen Mayberg and the Canadian surgeon Andres Lozano that the method got its breakthrough in psychiatry. It struck a sweet spot in the media when, in 2005, the two published the first study of deep brain stimulation for the treatment of severe chronic depression—the kind of depression, mind you, that does not respond to anything—not medicine, not combinations of medicine and psychotherapy, not electric shock. Yet suddenly, there were six patients on whom everyone had given up who got better.

At once, Helen Mayberg became a star and was introduced at conferences as "the woman who revived psychosurgery." Later, others jumped on the bandwagon, and now they are fighting about exactly where in the brain depressed patients should be stimulated. It is not just a skirmish between large egos but a feud about what depression really is. Is it at its core a psychic pain or, rather, an inability to feel pleasure?

"It's not my job as a neurologist to make people happy."

Helen Mayberg let her statement hang in the air between us before she continued.

"I liberate my patients from pain and counteract the progress of disease. I pull them up out of a hole and bring them from minus 10 to 0, but from there the responsibility is their own. They wake up to their own lives and to the question: Who am I?"

It was the end of February. The winter had been unusually harsh along the entire eastern seaboard in the United States, and even though the sun was now beaming down on Atlanta, it was cold. Emory University is a well-manicured private school with restrained classical architecture and winding asphalt paths between green areas. Helen Mayberg had a high profile in the academy, and her office spread out along the glass gable of the building. Physically, there was something elfin about her with her brown pageboy hair, which rested on the edges of a large pair of spectacles. She was a striking, diminutive figure. But she loomed large as soon as she began to speak. Her voice was deep and intense, and she let her words flow in a gentle stream that forever meandered in different directions.

"We had a hypothesis, we set up an experiment, we laid out the data, and now we have a method that works for a great many patients." She took a breath and lowered her voice half a tone. "But for me, it has always been about *understanding* depression."

Mayberg began her journey into the mechanisms of depression back in the 1980s—in a time when everything was all about biochemistry and transmitters. The brain was a chemical soup, and psychological symptoms were a question of "chemical imbalances." Schizophrenia was an imbalance in the dopamine system, and the serotonin hypothesis for depression was predominant. It claimed that this oppressive illness must be due to low levels of serotonin. The hypothesis was supported by the fact that certain antidepression medications increased the level of serotonin in the brain, but the theory did not have much else to back it up.

Then something happened to change the focus. There was a breakthrough in scanning techniques, and this meant, among other things, that you could look at the activity in living brains and compare what happened inside people with different conditions. During the 1990s, Mayberg began hunting for the circuits and networks on which depression played. Others were working in the same direction, and different groups could point out that there was something wrong in the limbic system as well as the prefrontal cortex. That is, both the emotional and the cognitive regions of the brain were involved. MRI scans of people suffering from depression revealed that certain areas were too active while others were too sluggish in relation to the normal, non-depressed control subjects with whom they were compared.

Soon, Mayberg focused on a little area of the cerebral cortex with a gnarly name, the *area subgenualis* or Brodmann area 25. It is the size of the outermost joint of an index finger, located near the base of the brain almost exactly behind the eye sockets. Here, it is connected to not only other parts of the cortex but to areas all over the brain—specifically, parts of the reward system and of the limbic system. That system is a collection of structures surrounding the thalamus encompassing such major players as the amygdala and the hippocampus and often referred to as the "emotional brain." All in all they are brain regions involved with our motivation, our experience of fear, our learning abilities and memory, libido, regulation of sleep, appetite—everything that is affected when you are clinically depressed.

"Area twenty-five proved to be smaller in depressed patients," Mayberg relates, adding that it also looked as though it were hyperactive. "At any rate, we could see that a treatment that worked for the depression also diminishes activity in area twenty-five."

At the same time, it was an area of the brain that we all activated when we thought of something sad, and the feeling that area 25 was a sort of "depression central" grew and grew as the studies multiplied. Mayberg was convinced that this must be the key—not just for understanding depression but also for treating those for whom nothing else worked. This small, tough core of patients who had not only fallen into a deep, black pit but were incapable of getting out again. These were the chronically ill for whom nothing helped, the kind of depressive patients who often wound up taking their own lives; it was this type of patient that, fifty years ago, were warehoused in state hospitals.

If only Mayberg could reach into their area 25!

And she could, with the help of a surgeon. Around the turn of the millennium, when she arrived at the University of Toronto, she met one of the institution's big stars, Andres Lozano. He had not only done deep brain stimulation on several hundred Parkinson's patients but was known as a researcher who was willing to take risks, who was eager to explore new territory. Here was something radical, and Lozano was more than intrigued. So it was simply a matter of recruiting patients. Over several months, the two partners spread the word, gave countless lectures to skeptical psychiatrists and, finally, began to get patients referred to them. One of them, a woman who had worked as a nurse before she became ill, was the first to sign up for the project. She had tried it all and did not expect an electrode to change anything. But why not give it a shot?

The operating theater was booked for May 13, 2003, and everything was made ready for the big test of Mayberg's hypothesis as well as her scientific narcissism.

"I felt the schism between my own curiosity and the patient," she said, holding both hands out from her body. "If something went wrong, it would be because *I* had asked a surgeon to do something on the basis of an idea."

But the surgeon patted her on the back and said that she, Helen, knew more about depression than anyone on the planet. Lozano himself was not the least bit in doubt that he could place the electrode in their patient's brain under very safe protocols.

"Ask yourself," he said to me, "if this were your sister, would you do it?"

Mayberg would, and they went ahead. The operation itself went by the book. The patient was told that there were no particular expectations.

"Nobody knew what would happen. So the patient was instructed to tell me absolutely everything she observed. Whether it seemed relevant to her or not."

The team began with their lowest-placed contact and 9 volts. Nothing happened. They turned up the voltage but still nothing happened. Then they went on to the next contact a half millimeter higher in the tissue. Even though they were only at 6 volts, the patient suddenly spoke. Were they doing something to her right then? she asked.

"Why do you think that? Tell me what you are feeling."

"A sudden feeling of great, great calm."

"What do you mean, calm?"

"It's hard to describe, like describing the difference between a smile and laughter. I suddenly sensed a sort of lift. I feel lighter. Like when it's been winter, and you have just had enough of the cold, and you go outside and discover the first little shoots and know that spring is finally coming."

Then, the electrode was turned off. And as soon as the electricity disappeared, the patient reported that her sense of springtime had vanished.

Now, years later, Mayberg pulled up her knitted sleeve and held her forearm out to me. She still got goose bumps when she talked about that first time. And when I asked about how she felt there in the operating room, she did not hesitate to admit that she was close to tears.

"There was a purity to the moment."

Later, it became clear that the reaction was not unique—other patients got the same "lift." For one, it was as if a dust cloud around her had disappeared, while another suddenly felt there were more colors and more light in the room. Once they had experienced this immediate effect, there was a good chance that their depressive symptoms would decrease over the first months after the operation. But the lasting effect came gradually, and it had nothing to do with euphoria or happiness.

"The patients are aware I have not given them anything but have *removed* something that was bothering them," said Mayberg. She liked analogies and offered me one. "It is like having one foot on the accelerator and one foot on the brake at the same time and, then, lifting your foot off the brake. Now, you can move."

This was the core of the Emory group's view of depression. They did not see it as a lack of anything positive—pleasure and joy—but as an active negative process. Neither did they believe you could just "inject positive" into a patient. Rather, you have to remove the constantly grinding negative activity.

This all sounded self-evident, but the stimulation did not work for everyone. Even though Helen Mayberg and her team of psychiatrists selected from a pool of more than 1,000 interested prospects a total of 27 patients who, to that point, had undergone the operation, they did not hit the jackpot every time. In the hands of Mayberg and the surgeon Robert Gross, it looked like 7 out of 10 patients experienced an effect that lasted beyond the first two years. Meaning that their symptoms of depression were at least halved. More recently, the statistics hold for 8 out of 11, of which many not only experience an improvement but what one would call normalization. Their condition is as it was before the illness struck them, and they have reentered the labor market.

Now, the team spent time dissecting its own work in order to find the source of the variation in efficacy. And some of it proved to be in the microscopic placement of the electrodes. "It is not just area 25 stimulation," as Mayberg put it. With tractography—that is, special scanning that visualizes the paths and connections of nerve fibers, they closely analyzed a group of patients and compared those who had a positive reaction with those who never felt any change. The result was published in 2014 in *Biological Psychiatry.* This paper claimed that it was not enough for the electrodes to hit area 25. The decisive factor was hitting a particular point where three bundles of nerves crossed. Presumably, it had to do with the fact that these three lines of communication had to fire up frontal regions of the brain simultaneously.

"With this understanding, we could go back and operate

again on six patients who had not reacted the first time. And now they experienced an effect," Mayberg said.

As if she had just thought of something, she pulled out a book, and the sight of it rattled me a bit. It was a small volume with a tan cover and white lettering, which said *Exploring the Mind-Brain Relationship* by Robert Heath. I had heard about Heath's monograph but could not get hold of it. It is a writing he self-published in 1996, three years before he died, with the ambition of explaining his scientific project and what had come out of it. The book was never for sale anywhere, and the publisher is no longer in business.

"It changes every time you read it," said Mayberg, tapping the cover with her index finger. "I heard a little about his work as a young student, but when I read it today, I am more cautious about judging the extent to which he was either a visionary or a lunatic. I can see him both ways, depending on how I look at it and through which prism I view it."

Heath ended up being labeled as ethically tarnished. However, Mayberg said, it may have had something to do with the tension between two fundamentally different views about how researchers should deal with the brain. On one hand, there are those who see the brain as the source of everything human and, therefore, the key to understanding everything human. On the other hand, there are those who believe that there are certain things you just should not fiddle around with.

Helen Mayberg herself had a run-in with the latter viewpoint. When her landmark paper was published in *Neuron* in 2005 and she gave interviews to the major newspapers, the blogosphere exploded with indignant scribblings. Doctors had crossed the line! This was the return of the lobotomy!

"The conflict arises every time science reaches a new frontier. And as soon as research has anything to do with the brain, there are people who get nervous that it can be used for *enhancement*."

That was my cue. I wanted to hear what Mayberg thought about joy, pleasure—hedonia. I know some groups treat depression by stimulating areas in the reward system and "injecting positive," as she somewhat mockingly calls it. This applied, in particular, to a duo from the University of Bonn—psychiatrist Thomas Schläpfer and surgeon Volker Coenen, who virtually churn out studies reporting impressive results.

A certain tension appeared in the room. Mayberg stressed several times that Schläpfer was a "friend and a colleague," but she also believed that he was in a strange competition with her. That it was as if he could not deal with the fact that she came first.

"There may be plenty of people who suffer from anhedonia and who might get quite a lot out of having an electrode placed in their reward system. But if you don't have psychological pain, I don't believe it's depression. If life just isn't good enough, it won't do anything for you to tone down area 25."

Mayberg related the story of a patient to me. This woman had an alcohol problem in the past and, after she had her electrodes installed, she went home and waited for them to give her a sense of intoxication or euphoria. She was completely paralyzed by her expectations, and Mayberg had to explain that there was nothing to wait for. The procedure had simply awakened the lady to the realities of her life. The symptoms of her disease were diminished, but she herself had to put something in their place if she wanted to fill her life.

"Our nervous system is set up to want more and to go beyond the boundaries we run into. You don't want just one pair of shoes, right? I fundamentally believe that you go into people's brains in order to *repair* something that is broken, but there is something strangely naïve about wanting to stimulate the brain's reward system. Ask any expert on addiction. You will wind up with people who demand more and more current."

I thought how different this view was from Robert Heath's. To him, pleasure was a cure in itself or certainly the key to a cure, whereas Helen Mayberg seemed to view hedonia as almost a bit suspect. Suddenly, she got up from the table and left the room, which remained quiet until she came back with a little picture. A white canvas on which the words "AREA 25" were written several times on top of each other with thick, black writing. It was a work by an American artist who felt inspired by Mayberg's endeavor. Somewhere out there, there was also a band that, for the same reason, called itself Area 25.

"I didn't ask for this, but it happened, and now I have to accept it and be the best possible ambassador for the entire field."

Helen Mayberg stroked the surface of the picture a couple of times, and I noticed her hands. They were work hands, with short nails, no polish.

"The lesson here is that you are a hero until you're not. That happened to Robert Heath. The best we can do is to learn from history in order not to repeat its mistakes."

• • •

My curiosity was piqued by what she had said about the reward system and addiction. I dove into the literature and found an article from 1986, which described a case of dependence on deep brain stimulation. The journal *Pain* had a so-called case history of a middle-aged American woman. In order to relieve insufferable chronic pain, she had a single electrode placed in a part of her thalamus on the right side. She was also given a self-stimulator, which she could use when the pain was too bad. She could even regulate the parameters of the current. She quickly discovered that there was something erotic about the stimulation, and it turned out that it was really good when she turned it up almost to full power and continued to push on her little button again and again.

In fact, it felt so good that the woman ignored all other discomforts. Several times, she developed atrial fibrillations due to the exaggerated stimulation, and over the next two years for all intents and purposes her life went to the dogs. Her husband and children did not interest her at all, and she often ignored personal needs and hygiene in favor of whole days spent on electrical self-stimulation. Finally, her family pressured her to seek help. At the local hospital, they ascertained, among other things, that the woman had developed an open sore on the finger she always used to adjust the current.

This sounded crazy, bizarre. The woman, who had previously had a serious alcohol problem, had been like one of the lab rats in the Olds and Milner experiment that neglected everything to tramp on their reward pedal. Or like people who will do anything for their next fix of heroin or another line of cocaine. But of course, it *would* be like that. If you can be brought into a state of ecstasy and addiction by drugs that affect the brain, you can, of course, bring about the same effect by direct electrical stimulation.

I had to go to Bonn to see what Schläpfer and Coenen were doing. I knew they were stimulating a "new" area of the brain they had explored in connection with depression. The medial forebrain bundle, as it is called, is located deep inside the brain, and scientists have long known that it plays a central part in the reward system of rodents. Coenen made the first good map of this area in humans. He discovered that it integrated the emotional limbic structures in the middle of the brain and also had projections to the prefrontal cortex. In addition, new studies showed that the bundle's nerve fibers were thinner in people suffering from anhedonia.

The two Germans now claimed that stimulation of the medial forebrain bundle worked faster against depressive symptoms than anything else.

"Good morning, Frau B. It's time. We're here to take you down to prep for the operation."

Christina Switala bent over the hospital bed and looked into Frau B's eyes as she took her hand. "You've been waiting for a long time for this. Remember that you can ask about anything along the way," she said. Switala was a psychologist, and she kept close tabs on all the electrode patients in Bonn. She spoke to them before, during, and after the operation, and she collected the data during their long course of treatment. She had dark hair and porcelain skin and was surprisingly young—in her early thirties, I guessed—but was one of those people who naturally exude confidence. Frau B nodded weakly from her pillow.

The patient, who was sixty-two years old, had recurring depression most of her adult life, but for the past almost three years, her condition had been deep and chronic and impossible to alleviate. Lying flat on her back as her bed was guided down the broad white hallways, she seemed, more than anything, absent. Reality only set in when she arrived for the prep and noticed the stainless-steel instruments laid out and the stereotactic frame that was to be attached to her cranium. Then she became nervous and wept loudly.

"What's this? Tears?" asked a young man who suddenly stepped through a door. "There's no call for that, and you have no time for that kind of thing today. Now, we just need to get this fixed up."

He had wide sideburns and a ponytail, and the embroidered letters on the pocket of his lab coat read *Dr. V. Coenen*. At first, I could hardly believe it. Neurosurgeon Volker Coenen had a hefty professional reputation, and I had always imagined an older German professor of the more formal type. This guy was only in his early forties and surprisingly informal. He chatted away enthusiastically as he went about shaving the hair off the snuffling woman in the bed. When her blond locks fell from the front half of her scalp, she began to cry, and when, a moment later, Coenen injected her forehead with a local anesthetic, the crying turned into a loud yowl.

"Ow, ow, ow, it hurts too much! If I had known it would be like this, I never would have agreed to the operation," whined Frau B.

"And if I had known this would happen, I'd have put you under full anesthesia," Coenen mumbled behind the head-board. He turned and whispered to me in English that this was very unusual. Patients were normally very cooperative. Finally, it became too much for him.

"Frau B. We have a big job in front of us, and I can't work with a patient who is sobbing and needs her hand held. If this is to work, you have to be strong and pull yourself together. Do you understand?"

After a bit more snuffling, Frau B finally succeeded in holding back her tears; a few long minutes later, the stereotactic frame was screwed to the bone in her forehead and neck. The patient's petrified face was fixed between two bars. The whole thing looked like an advanced form of medieval torture. Volker Coenen was satisfied with the result and took hold of the bed to move his patient to a CT scan before things got going.

"We're ready."

Ever since I got the invitation from Coenen and Schläpfer, I had been looking forward like a little kid to seeing actual brain surgery. I had seen human brains in formaldehyde and even held one in my hand, which certainly had its charms. But there was something almost magical about being able to look inside the quivering organ—the living self—and actually touch it.

The atmosphere in the operating theater itself was almost cozy. The entire back wall of the square room consisted of large panoramic panes, and a soft, pleasant light filtered through the beech forest on the other side of the glass. We were just short of a dozen people in there. Everyone was dressed in the same shapeless, light-blue operating gowns with matching surgical caps on our heads and masks over our faces. A sort of medical burka.

At the center of the room was Frau B, strapped down and sitting up in her bed with her back toward the windows. A large sheet of transparent plastic was stretched around her head. The area behind the plastic was Volker Coenen's workspace, where instruments, wires, cotton, and various other accessories are laid out on trays in accordance with some ingenious system. Every little piece was sterile and guarded zealously by a tall male nurse with restless hands in rubber gloves.

"If you can't stand the sight of blood, just sit up against the wall. We can't have anyone collapsing and getting hurt," Coenen said, speaking to me. Fifteen seconds later, he began to drill. I don't know how it feels to have holes the size of a nickel drilled into the top of your skull while you're wide-awake; but, from the outside, it doesn't sound very pleasant. The drill looked like an ordinary craftsman's model, and Coenen put his weight behind the work. The sound was like when you drill into a concrete slab. A yellowish dust sifted out from the edge of the drill, mixed with a little blood. Skull dust, I thought to myself.

"Frau B is the most treatment-resistant patient I've ever had," someone said quietly from behind me. It was Thomas Schläpfer, the psychiatrist behind the project and Volker Coenen's inseparable partner, who had crept in to check up on his patient. He was tall enough to tower over most people and sufficiently corpulent to seem imposing. In his blue outfit, he looked like a large, serene teddy bear with black steel-rim glasses.

In the meantime, Coenen worked inside the patient's head. The first thing he did was to lower two thin, rigid tubes down into the brain tissue; then it was time to send a test electrode through the tube. With that, Coenen measured his way to the perfect place to stimulate. From scans of Frau B's brain, he had coordinates and a good overview of the area, but he still had to make sure that he hit the sweet spot. Coenen came around to the other side of the plastic, turned on a computer, and settled down in front of the screen. At one point, he looked satisfied with the signal and gave instructions to turn on the current. As the first electrical pulse streamed into the patient, Coenen turned toward the blood-pressure monitor hanging above the foot of the bed and waited for the slight rise that was the sign he was in the right place. *Blip*, said the apparatus.

"There. That's it."

Schläpfer asked me to take note of the patient's eyes, and something was actually happening. They suddenly opened a bit, and she looked around the room, whereas before she just stared blankly into space. It was a subtle sign, and more signs gradually appeared. After a couple of minutes, Frau B began for the first time during the hour-long procedure to take part in conversation around her. She reported in with a couple of questions and, at one point, said something that resembled a little joke.

"Did you hear that?" Volker Coenen nudged me. "She said she *hopes* the treatment will work. This is the first time she has ever spoken to us about hope."

• • •

While the holes in Frau B's cranium were being closed and her scalp sewn nicely back together, I followed Schläpfer out into the hallway and asked about the innermost nature of depression. What did he and Coenen think about Helen Mayberg's ideas about psychological pain needing to be extinguished, as opposed to countering anhedonia?

The big man sighed, paused for a moment, and answered by way of an anecdote from when he was studying at Johns Hopkins in Baltimore. On one of the regular hospital rounds, the old head of psychiatry at the university pointed at him and asked him to name the symptoms of depression. The dutiful Swiss student stood up straight and began to recite the nine symptoms from the textbook, when the old man interrupted.

"No, no, young Schläpfer. There is only one symptom, and it has to do with *pleasure*. Ask the patient what he gives him pleasure, and he will tell you: *nothing*."

Young Schläpfer thought about his superior's remark and actually began to ask his patients questions. He still does. Today, he believes that anhedonia is the central symptom while everything else, including psychological pain, is something that comes in addition to that. It is only when their anhedonia abates that people suffering from depression feel better. And this is not strange, because desire and enjoyment are driving engines and a key to many of our cognitive processes. Desire pushes, so to speak, all the other systems and even makes it possible to have motivated behavior and to work toward a goal.

"I am familiar with Helen's attitude towards the reward system," said Schläpfer with his slow diction. "But I would like to stress that we have never seen hypomania in the patients we stimulate in the medial forebrain bundle. If we overstimulate and turn the current up too high, the worst reaction we have seen is that people get a tingling sensation as if they'd had too much coffee."

That sounded harmless enough. But you *could* achieve euphoria or something that looked like it if you chose another spot in the brain—for example, the nucleus accumbens, which some groups use not only for the treatment of depression but also to treat OCD, anorexia, and overeating. And what about people who demanded euphoria, as the young man Dr. Schläpfer himself has written about?

When deep brain stimulation is no longer experimental but an approved standard treatment, anyone can take their stimulator and pay a visit to a doctor willing to set it right where they want it. Hypomania be damned. And why not leave it up to the patients how they would like their brain tuned? In principle, is there no difference between improving something from –5 to 0 and from 0 to 5?

Schläpfer called this a fair point and did not see anything necessarily unethical about making whatever enhancement the patient wanted. He also predicted that we will, presumably, reach a point where patients may choose to have brain stimulation done to improve selected mental capacities.

"It will undoubtedly be possible to enhance various abilities electronically," he said. "But probably at the cost of something else."

Schläpfer himself had actually seen cognitive enhancement

in his electrode patients. All deep brain stimulation groups do thorough neuropsychological tests of their patients before and after stimulation, but the Bonn folks had a study showing that every participant did better cognitively after the operation. It was not because their depression had disappeared and thus no longer tapped into their intellectual powers. All of them showed improvements regardless of whether their depression had changed or not, and the improvement was seen across a number of cognitive domains, from linguistic ability to complex problem-solving. An unexplained effect of stimulation and apparently independent of mood lifting.

Others had seen improvements in memory. Andres Lozano at the University of Toronto told of a very overweight man who came in for stimulation of his thalamus, which regulates appetite. He was on the operating table and, when the current was turned on, he was flooded by memories. Powerful, vivid images of long-ago-forgotten episodes in his life poured out and, when the doctors later tested him, he could remember far better when the electrode was on than when it was off.

"It was that patient who inspired Lozano to do his first experiments with deep brain stimulation on Alzheimer patients," Schläpfer remarked. But I wondered whether the inspiration might go much further than that. Can it be imagined that, over time, neurosurgery will develop along the same lines as plastic surgery?

Plastic surgery was developed in the wake of the First World War in attempts to repair the shattered bodies of war veterans and to treat the abnormal and deformed, which for a long time was the field's only goal. But as we well know, this changed. Today, cosmetic plastic surgery is a huge global business. Highly specialized professionals enhance noses, breasts, even labia, in accordance with the fashion of the day and the customer's varying wishes.

The question is whether deep brain stimulation might open the same wide floodgates for neurosurgery. We are living in a time in which we see ourselves as biological machines, and through that lens it can be difficult to see anything wrong with upgrading your inborn hardware. Because it is only that—hardware.

"No, why not? We also exist in a market and, if there is a demand, a supply will arise. Personally, I wouldn't call such people doctors . . ." said Schläpfer, referring to hypothetical neurosurgeons performing enhancements. But before that problem would materialize, he was worried about something else entirely. What bothered him were the clinical and placebo-controlled trials of deep brain stimulation, which had been disappointing. In the United States, the major producer of stimulators, Medtronic, had done a trial of nucleus-accumbens stimulation for depression, but it failed to demonstrate a statistically robust effect. In Canada, St. Jude Medical was responsible for the so-called BROADEN trial of area 25 for depression. It was suspended halfway through because an analysis of the interim results revealed that it was not likely to demonstrate any improvement in the group being treated.

The failures were not due to the method, said Schläpfer, but to commercial haste. The producers of electronic equipment had been far too quick in trying to get treatments approved and disseminated to the masses of waiting patients. That was why they were throwing themselves into trials before the technique was clear. Just because one research team developed a technique that worked in their hands, it was by no means certain you could just send a description to any neurosurgeon, ask him or her to do the operation, and get the same results. Even when everyone involved was trying to proceed as scientifically as possible, it was as if there were also a sort of art to it. Helen Mayberg told me that she had reached the point where she just *knew* which patients would show an effect by meeting and talking to them. It was practical experience, which was indefinable and could not be laid out in a set of instructions.

"The danger is," said Schläpfer, examining his hands in his lap, "that mainstream psychiatry will lose interest, no longer believe in the method, and write it off entirely. It could kill deep brain stimulation."

Helen Mayberg's parting words were that the present must avoid the mistakes of the past. But what were those mistakes? What was it that killed electrode stimulation the first time around? Robert Heath was not alone in this kind of experimentation. In the 1960s in particular, there were groups in both the United States and Europe—not to mention one Russian scientist—exploring the possibilities.

At Yale University, the flamboyant, Spanish-born neurophysiologist José Delgado engaged in a series of animal experiments to identify brain mechanisms behind different types of behavior. Delgado discovered a method of implanting electrodes that could be activated by radio waves and did not require cords and wires. He demonstrated how effective it could be with a spectacular experiment on a bull. In 1964, the tall man, carrying no other equipment than a radio transmitter, walked into the arena with a battle-trained beast at a bull farm in Cordoba. When the bull lowered its head and began running directly at him, Delgado allowed it to come within a few meters of him before he activated his apparatus. At once, the animal stopped and stood still. By activating an area of the brain called the caudate nucleus, which is involved in the control of movement, he interrupted the motor program that was propelling the animal forward.

Delgado tried out the same cordless technology in the late 1960s on a handful of humans. Along with Frank Ervin, who at that point was affiliated with Harvard, he experimented by stimulating violent patients with psychomotor epilepsy.

In Oslo, the Norwegian psychiatrist Carl Wilhelm Sem-Jacobsen, who had visited Heath in New Orleans, also brought brain stimulation home with him. There, he studied the stimulation of psychotic patients and reported that their symptoms had been significantly diminished. But unlike Heath, Sem-Jacobsen used his studies to identify brain areas with problematic activities and then burned them away. Good old-fashioned psychosurgery, in other words.

In the same way, both a Russian and a French group used electrodes on patients with Parkinson's and epilepsy. They identified brain areas that were the cause of tremors and unleashed epileptic fits, and they removed them.

In 1966, there was a world congress for psychiatrists in Madrid, and those particularly interested could attend a symposium on the use of electronic equipment in psychiatry. This involved computers, which people were slowly starting to develop and seemed to have such a great future, but it was just as much about the implantation of electrodes. The Tulane group and a handful of other teams presented their work, and the lectures were later collected in a volume edited by two prominent American psychiatrists—Nathan Kline and Eugene Laska from New York. Their foreword explained what they thought about the new developments:

"The editors and, presumably, everyone else whose work is represented in this book have great confidence that, over the next few years, electronic equipment will bring about great changes in psychiatric diagnosis, therapy, and research," they wrote. And they ended with optimism by predicting that "psycho-electronics is undoubtedly an important part of our future."

But in the 1970s, it was as if this whole branch had disappeared. This was the decade in which the great pendulum of society swung far to the left. The youth culture flourished, and authority was in disrepute. In the eyes of many, psychiatry looked like something out of *One Flew over the Cuckoo's Nest*. The literature shows that several places in the United States and Europe were stimulating patients with chronic pain, and one of the more prominent practitioners was Donald Richardson, a neurosurgeon who was with Heath at the very beginning and saw, among other things, terminal cancer patients experience total pain relief.

But reports about psychiatric patients disappear at this time. Why?

CHAPTER 5

The Times They Are A-Changin'

Stop the mind control!

No more crazy experiments!

Hands off, Dr. Heath!

It was a hot day in May. The year was 1972, and a group of angry young people had taken up a position outside an expensive conference hotel on the outskirts of New Orleans's French Quarter. They pumped their handmade protest signs up and down while some of the men shouted slogans against Robert Heath and his "crazy experiments." Inside the hotel's chilly rooms was a group of the nation's leading EEG specialists, who had assembled for their annual conference and did not entirely grasp what was going on.

Most of the demonstrators were from the local Medical Committee for Human Rights. They were nurses, paramedics, porters—there were even a couple of young doctors—all called together by Todd Ochs, who drove an ambulance for Charity

Hospital and also helped out at a free clinic for the city's less well off. Ochs himself was headed toward the study of medicine and had thought of becoming a pediatrician. But he was deeply suspicious of parts of the system and especially psychiatry, with its ossified old geezers who believed they could do whatever they wanted to defenseless patients. Just look at this Robert Heath: He was deemed a god on Parnassus there in the city. Ochs had never met the man face-to-face, but from everything he had heard, Heath was clearly the devil himself clad in a white lab coat.

The last straw for Ochs had been a recent article in the *Journal of Behavioral Therapy and Experimental Psychiatry.* Its description of patient B-19, who was supposed to have his homosexuality cured with the help of a bunch of electrodes and a prostitute, had struck him in his moral solar plexus. Medical research was supposed to help people with their illness, not to "correct" their behavior. The thoroughly clinical description made him sick: The man who masturbated in front of porn movies while the scientists were studying him, and the whole bizarre scene with its prearranged sexual intercourse were presented as if it were an appendicitis operation that had taken place on just any ordinary Tuesday morning. It was sick.

"What kind of 'scientist' would even think of this?"

Ochs had also discovered that the young man, Patient B-19, had originally been caught by the police in possession of drugs in the nearby town of Lake Charles. He agreed to hospitalization and observation at Charity in order to avoid being charged. In addition, there were rumors among the personnel at Charity that nurses sometimes hid female patients when Heath's "lackeys" were "on the prowl for victims."

Todd Ochs was incensed and felt he had to do something. He distributed a flyer throughout the area, and people had responded. It already felt like something of a triumph when one of the otherwise smug professors peeked nervously through the curtains in the lobby from time to time. But the protest was not actually directed at the doctors—they were beyond rehabilitation. It was for the public. If anything were to be done, they had to call on the people, get some media attention, and reach the folks who had political influence.

They never even got a glimpse of Robert Heath. He was supposed to have presented new data to his colleagues but got wind of the trouble and stayed away. He also canceled the meeting that had been set with the young people and their human rights group. As if he were committing an offense against human rights! He had been willing to explain his work to them, but if they decided to behave like thugs and rabble and refused to play according to the ordinary rules of professional conduct, they could go suck eggs. The idea that they would stand there in the street shouting his name made him furious. Heath saw them as a bunch of longhaired ignoramuses and shared his view with his closest staff.

But there was something else as well. He was not entirely sure why they were so upset about B-19 and his treatment. This was a published scientific study that had been peer-reviewed: a classic case of behavioral conditioning and an attempt to redirect neurotic behavior with which the patient had big problems. With his mixture of schizophrenia and deep personality disorders, B-19 was not exactly an easy patient, and they had gone to a lot of trouble to help him. For months, he had been enrolled in the treatment program, and the group had used countless man-hours talking to him and designing the electrode treatment. And for a time, it actually worked. After he left the hospital, B-19 had a long-term relationship with a woman.

Certainly, Heath may have had a feeling that questions might be raised, but that was why he was so careful to get consents and approvals all the way down the line. He went all the way to District Attorney Jim Garrison to get approval for the prostitute, and there were no problems—the DA was interested and even helped them find a suitable woman. The porn was also negotiated with Garrison and brought in from the downtown police warehouse for confiscated illegal goods. Hot stuff, some of it, with three-ways and people who were into all manner of perversion. This was exactly the point, to check the measurements to see precisely what turned B-19 on and whether they could influence his preferences.

In fact, this was an advanced and deeply interesting experiment that pointed the way forward in many ways. He had talked about it several times in the media. He had explained how direct stimulation of the brain's pleasure center presumably had the greatest potential for affecting neurotic behavior. As he had explained two years earlier in the popular journal *Medical World*:

> The neurotic patient is maladjusted because he has developed a mistaken fear. By giving the fearful and phobic patient the experience of a pleasurable feeling when he

> encounters the object of his fear, the procedure has the greatest potential for correcting the maladjusted pattern.

At that time, no one lifted an eyebrow. Not even when he, as late as the year before, laid out his work with B-19 in *Newsweek*—it was the front-page story in the magazine, which was enthusiastic.

And now there were angry demonstrators. What was their problem? God knows, he had nothing personal against homosexuals. New Orleans was notoriously full of them—they said every third resident in the French Quarter was gay—and that included some of his own medical students. Not that they'd said anything, but he knew. Take his senior resident for many years, Jim Eaton, as an example. Eaton, with whom he'd had a close mentor relationship for many years, was a good friend, and Heath had always known that the young doctor from Mississippi preferred men. And it was said that all the physicians at Charity's emergency room were homosexuals. But you could not get around the fact that it was a neurotic ailment that was listed in the psychiatric diagnosis manual—and many homosexuals badly wanted to be rid of it.

Heath betrayed nothing of his feelings outside his office and in Irene's company, and he continued to behave as he usually did. The lab technicians and the shrinking violets on the administrative staff still had to put up with the fact that he could behave "eccentrically," as some of the more seasoned personnel called it. He might walk past a desk and drop a pair of lace panties, saying things like "look what I found in the back seat of my car." He also continued to park his car around the city without any consideration for the rules of the road or respect for traffic lanes. The police knew his car, and it was understood that he would not be ticketed.

But beneath that well-polished exterior, some of his closest colleagues thought they saw a small, budding insecurity. Was it an unpleasant sensation of suddenly not understanding the world? It was as if something had happened that he could not have seen coming and that he did not know how to react to. Something had tipped over, and a new mood or zeitgeist was spreading out like a sticky goo around him.

The year after the demonstration in New Orleans, he was called to Washington, DC, to explain himself. There was a much bigger movement afoot. Psychiatry itself was being scrutinized under a harsh light all over the country, and the area Robert Heath was involved in had become the lightning rod. In 1973, he received a subpoena from the US Senate to a subcommittee of the Committee on Labor and Public Welfare, which wanted him to testify at a critical hearing about human medical experiments.

The man who had whipped up the hysteria was himself a psychiatrist. Peter Breggin practiced in Washington, DC, and he had a very big thorn in his side about psychosurgery. Personally, he was in the anti-psychiatry camp, which claimed that psychological diseases were not illnesses at all but natural reactions to the pressures of society and environment. As a consequence, patients were not to be treated by interventions in the brain.

Breggin despised the old lobotomists who, in the 1940s and

1950s, had been responsible for thousands of procedures that destroyed lives. Now he'd discovered that some of those butchers and their disciples were active again. They were planning a big comeback for their despicable practices, he believed, a sort of second wave of psychosurgery. He saw himself as a bulwark against it and mounted a direct attack with a paper entitled "The Return of Lobotomy and Psychosurgery." At first, no one in the media would publish this long tract by an unknown writer, but when, through a political connection, Breggin had his text inserted into the Congressional Record, there was interest. The Associated Press dispatched a telegram, and his success was made.

The matter was set for a hearing, and Breggin called urgently for a law banning psychosurgery—at least, when it was not a properly approved treatment but purely experimental. Some of the villains he pointed to were Heath's old protégé, Frank Ervin, and his new partner, neurosurgeon Vernon Mark. The two were affiliated with the prestigious institution of Harvard and practiced at nearby Massachusetts General Hospital, but what they did roused Breggin and others up out of their chairs. Ervin had pursued his interest in the biology of violence. He and Mark were treating a group of patients with psychomotor epilepsy and uncontrollable rage by removing small pieces of their amygdala. They received public funding for their work, and according to Breggin, both the researchers and the generous people at the National Institute of Mental Health presented results far rosier than they actually were.

For Breggin, there was no difference between permanently removing tissue from the brain and poking electrodes into it.

However advanced the electronics were, it was still an absurd method to subject people to, and there was no way it could be called therapy. This was nothing but experimentation, and it was dangerous. He named Heath as the leader in the field and referred to a "disturbing picture" in the magazine *Medical World*, which showed one of the stimulated patients.

"Heath has the record with 125 electrodes implanted at one time—it is a brain that has been made into a human pin cushion."

At the hearing, the committee chairperson Edward Kennedy, senator from Massachusetts, spearheaded the questioning and, after a long tirade from Breggin, he welcomed Robert Heath.

The impeccably dressed professor duly thanked the senator and jumped directly into an argument for experimentation as a precondition for developing new treatments. His own experiments, which Breggin and others found bizarre, were in support of treatment. They simply had to use these invasive—though gentle—electrodes and glass needles to learn things about the brain.

"With these tools, we have been successful in building a meaningful bridge between mental activity and physical activity in the brain, the organ of behavior."

Then, Heath adroitly built a bridge to the question of ethics, stressing that they were on top of it. At Tulane, they had set up a committee of doctors, lawyers, and even clergymen who were to approve all experiments. Moreover, the research subjects and their families had to provide informed consent. They even talked with a doctor who was not himself involved in the project.

"You believe this is terribly important?" Kennedy asked, looking down at Heath in the witness chair.

"Terribly important," he replied. Then he turned the ethical question upside down for the senator. Is it defensible for a doctor to desist from a method of treatment that, certainly, bears risks but also has the chance for improving the patient's life if the alternative is lifelong confinement to an institution?

"For example, one of our patients had received psychiatric treatment for years without effect when we began to implant electrodes. She and her family offered to testify today because she is doing so much better."

"Very good," Kennedy mumbled. "Perhaps we could have the films?"

Certainly. Heath had brought along a bit of everything and showed the flabbergasted assembly how, after the patient Joe received a few volts of stimulation, he went berserk. He shouted that he desperately wanted to kill someone, whom Heath identified as the surgeon, Dr. Llewellyn, who was just outside the frame—but only until they turned off the current. Then they could speak to him, as he made associations to episodes much earlier in his life, when hotheadedness and rage ran away with him. By identifying the pain circuit, they could learn to treat the discomfort, Heath explained, starting the next clip: a psychotic man in a straightjacket (due to a recent fit) having his pleasure center stimulated and becoming calm.

The members of the commission whispered to one another, and Kennedy leaned into the microphone and asked what the implications of all this were—five or ten years into the future.

"No matter how we treat a patient with behavioral disorders, we have to be cognizant of the role of emotion, painful emotion as well as pleasureable. This is important in the whole learning process in normal human subjects.

"And as for the sick—neurotics and the psychotic—there is an abnormal link between feelings and behavior.

"If we are to understand and treat behavioral disorders, it is crucial that we establish the brain mechanisms underlying basic behavioral phenomena of feelings and emotion."

"What you are really talking about is controlling behavior?"

Heath replied that, as a doctor, he was not interested in controlling behavior, but all tools could, of course, be used and misused.

"Wouldn't you also be able to treat people other than the sick, i.e., normal people," asked Kennedy. "And shouldn't we be worried about mass application?"

It amazed Heath that the senators were so terribly sensitive to the allegations of mind control that were bandied about. It was a completely muddle-headed debate. He tried to put it in clear terms: There was no prospect that masses of people would get electrodes implanted in their brains for fun. It was too expensive and too impractical. And the fact of the matter was that there were far easier ways of manipulating people's pleasure centers. If anything exercised control over the will and manipulated people's behavior, it was narcotics, which young people had begun to take a lot of. Particularly marijuana, which he had recently studied on monkeys and ascertained damage to their brains. Everyone seemed to think that it was a harmless recreational pastime, but he concluded his testimony by warning against drugs in general.

"I think they will continue to cause problems if their use cannot be controlled effectively."

After an entire day's proceedings and hearings, the witnesses went home to their universities, and the committee appointed a commission to examine the problem and provide a recommendation. The National Commission for the Protection of Human Subjects of Biomedical and Behavioral Research worked on it for the next four years.

Back home in New Orleans, journalist Bill Rushton had ruminated about things long enough. He had no doubt that these doctors with their need to control people's behavior had to be opposed.

"The national struggle against the tyranny of psychosurgery has grown in strength over time," he wrote in the fall of 1974 for the alternative left-wing magazine the *Courier*. There, he provided his own contribution to the struggle in the form of a long exposé article entitled "The Mysterious Experiments of Dr. Heath, in Which We Wonder Who Is Crazy & Who Is Sane."

Rushton was active in the city's homosexual community and a close friend of Todd Ochs. Since the demonstration two years earlier, the two friends had been egging each other on, and Rushton wanted to do his part to stop Heath's abuse of brain surgery. The journalist wanted to open people's eyes as to what was going on at the otherwise untouchable Tulane and expose the celebrated professor for what he really was.

"Who *is* this mystical Dr. Heath?" asked Rushton, calling him "media-shy." "Where does he come from, and what has he been doing all these years?"

Like Ochs, Rushton had never met or spoken to Heath, nor had he interviewed anyone from Tulane. On the other hand, he had trawled through quite a few of Heath's scientific publications all the way back to the beginning, and by compiling quotations from them, he effectively described the research at Tulane as a chamber of horrors, drawing a picture of Robert Heath as a sort of counterpart to the Nazi doctor Josef Mengele.

Rushton got from Ochs one of the consent forms that patients and their families had signed, and he argued that, in reality, it was a carte blanche to subject patients to anything at all. Once again, there was indignation about the control of people's conduct. In particular, the descriptions of patients B-7 and B-10 with their self-stimulators hanging on their belts aroused the ire of the journalist. They are "zombies with transistors," as he wrote. And what should one think about the experiments with women who had chemicals injected directly into their brains, so Heath and his men could study their orgasms?

The *Courier* was more or less an underground newspaper, but it still reached Heath and his colleagues. People at the university were shocked on their boss's behalf. This was something much different from the kindly disposed journalists at the *Times-Picayune* with whom the press office at Tulane was used to dealing and fed with interesting new results, which they always received graciously. These were journalists "with whom we know we can work," as the press office formulated it.

Don Gallant would point out to anyone who wanted to listen

how the *Courier* article was full of nonsense and point to a paragraph that claimed Heath ruled over a special ward of 133 patients he could experiment on. "Those people are *my* patients, and they have never even seen Bob Heath!" The ward was devoted to regular, controlled drug trials—of which Gallant himself was in charge—and every dot and jiggle of the new guidelines for information and consent were followed.

What was all this about Heath and his work having "been hidden away at Tulane for a quarter of a century," as Rushton claimed? It had never been secret what the ward was up to. Bob, in particular, had made himself available to pretty much anyone in the media—to such a degree that his colleagues sometimes believed it was a bit too much.

Of course, the Heath family also read the *Courier* article, and things did not go gentle into that good night for them. Heath's son, Robert Jr., who was enrolled at Tulane, did not want to show his face around campus. Everyone knew who he was, and suddenly a lot of people wanted to tell him what they thought about his father. Rob had nothing to do with his father's experiments and preferred not to talk about the matter. His older sister was forced to. Her husband, who was also a doctor, mocked his father-in-law for his mad enterprise—couldn't she see that the old man was that sort of guy who liked to play God and make a big deal about being indifferent to the opinions of others?

Of course, the crux of the matter was neither Robert Heath nor his controversial patient B-19. Rather, the story about B-19 and electrode treatments generally was the perfect crystallization of a more general clash between different views about what a human being is. Deep down, it was about whether human behavior, as it happens to manifest itself, should be manipulated and, if so, by whom. It was a conflict that was smoldering throughout the whole youth rebellion of the 1960s, and it broke out into flames during the mid-1970s.

One of the people who put it most clearly and without a lot of qualifications was Robert Heath's colleague José Delgado, the Yale professor. The flamboyant, Spanish-born neurologist had conducted brain-stimulation experiments along the same lines as Robert Heath. Unlike Heath, however, Delgado was not so interested in treating patients and was more focused on understanding fundamental brain mechanisms. His experiments involved cats and monkeys for the most part, along with his famous performance in the bull ring, but there was no doubt about the true goal. The knowledge about the brain was ultimately to be applied in humans, and it was to be used to transform society itself.

Delgado described his vision in a book called *Physical Control of the Mind—Toward a Psychocivilized Society*, which was published in 1970. There he sketched out the contours of a society in which technology and mental training were to be used in a targeted way to bring primitive, instinct-driven human nature under control. The great and decisive question for Delgado was "How are we to fashion behavior in the present and in the coming decades?"

Left to ourselves, we are quite simply too primitive to deal with the modern society we have constructed.

"The direction of the colossal forces discovered by man requires the development of mental qualities able to apply intelligence not only to the domination of nature but also to the civilization of the human psyche," he wrote.

José Delgado spoke warmly in favor of a sort of neuro-utopia. But this was a place that seemed terrifying to quite a few people. Just how terrifying may be seen in a 1974 film called *The Terminal Man*, based on Michael Crichton's science-fiction novel of the same name. The story was popular, and it effectively equated electrodes in the brain to dangerous mind control: Harry Benson is a young computer specialist who develops psychomotor epilepsy after an automobile accident. Every so often he suffers a fit in which he attacks and abuses people without remembering anything later. But something can be done. Benson comes under the knife of two visionary surgeons, who place forty electrodes in his brain and a minicomputer to control his fits. However, Benson discovers that the fits give him a rush of pleasure, and he learns to induce them instead of keeping them away. He develops into a pure psychotic murderer, and he is on the verge of killing his psychiatrist before she finally shoots and unintentionally kills him.

Crichton did not pluck his story out of the thin blue air or his own well-developed imagination. His inspiration came from his internship with Frank Ervin at Boston City Hospital at the end of the 1960s. There, he heard in detail about Ervin and the surgeon Vernon Mark and their treatment of violent epileptic patients. Many of them were also described in Ervin and Mark's book *Violence and the Brain*, which was published in 1970 and quickly brought the two researchers into controversy. It was the first book to cast light on how violence comes from within and was, so to speak, an activity in specific parts of the brain.

One of the clinical cases went into the book under the cover name of Thomas R. He was a thirty-four-year-old engineer with a career, important patents, and serious problems with his temperament. Fourteen years earlier while he was in the army, Thomas suffered a bleeding ulcer and went into shock—his blood pressure fell, and he went into a coma. He was in it for so long that his brain was damaged. Later, at regular intervals, he would lose consciousness and explode into almost psychotic violent fits.

Sometimes the violence was vented on friends, but most often it was aimed at his wife and children. A fit began with pain in his face. Sometimes, he drooled and smacked his lips. The next phase was that his thoughts were sucked down into a paranoid hole in which it looked like everyone was after him and his wife was cheating on him with a neighbor. She defended herself, but it rang hollow. Finally, he would go after her—typically picking her up and slamming her against the wall. He might do the same to his children if they happened to get in the way. Afterward, he was always wretched and bewildered. "Why did I do that?" he complained, and had to be consoled. Seven years of talk therapy with a psychiatrist, however, had had no effect.

Ervin, by contrast, placed measuring electrodes several places in Thomas's temporal lobes, and over ten weeks they measured and stimulated his brain with various frequencies and discovered that there was a particular place in his amygdala that robbed him of control. When they stimulated there, he would flip out. If they went four millimeters to one side, they would achieve the opposite effect. He relaxed and described it as if he were on the drug Demerol. "I am floating on a cloud and feel like I'm looking at the world through a TV screen."

Every day, they stimulated a spot in the lateral amygdala, and it kept Thomas free of fits for three months. It was the same as with Heath's patients—pain and pleasure kept each other at bay. Ervin suggested removing the epileptic focus by allowing the electrode to burn away a small bit of tissue around the tip. An easy, simple procedure. The patient at first said, Yes, please, but later went into a rage, screaming that nobody was going to destroy any part of his brain. It took weeks of persistent persuasion before Ervin and Mark finally got permission to carry out their operation. In their book, they concluded their account by stating that, in the four years since the intervention, Thomas R had sporadic epileptic episodes but never a fit of rage.

Crichton, who even in his student days had sensed that his future lay elsewhere than practical medicine, could recognize a good story when he saw it. But it knocked Frank Ervin for a loop.

When Harry Benson was blown up on the big screen, a lawsuit was commenced by an older couple who claimed that Ervin's Thomas R and Crichton's Harry Benson were actually their son, Leonard A. Kille. And neither the scientists nor the author had permission to write about him.

Ervin and Mark had been invited to establish the Center for Human Behavior at UCLA. Now, suddenly, there were demonstrators outside, and it was not a joke. The protestors aggressively tried to intimidate anyone presumptuous enough to attempt to enter the building to work, even beating them. One of those who received some lumps that day was a young, ambitious student by the name of Helen Mayberg.

It all came to a head when the grant that had already been given by the Justice Department was withdrawn. The center was dissolved, and Frank Ervin simply decided to give up work with human patients. He went to Canada, accepting a professorship at Montreal's McGill University. There he looked around for opportunities for doing experiments with the next best thing—namely, monkeys. As he said, "Actual patients—those we would like to help—we can give up on. From now on, we'll be curing schizophrenia in mice."

In 1978, the so-called *Belmont Report on Psychosurgery* was duly published after four years of work. It concluded on the basis of the evidence collected that psychosurgery actually *did* help groups of patients with psychiatric ailments. Peter Breggin and the other critics did not get the ban they had hoped for. But that did not matter because so much mud had been thrown at the entire field in the process that, for all practical purposes, it was dead. Neurosurgeons wisely kept their mitts away from psychiatric patients in order to avoid any problems. And at Tulane, troubles continued to wash over Robert Heath, but from a slightly different front. In the spring of 1978, he got a call from an eighteen-year-old law student who wanted to dig up the remote past.

"My name is Douglas Nadjari, and I'm a reporter for the

student newspaper the *Tulane Hullabaloo.* I'd like to ask you a few questions about your collaboration with the CIA, Dr. Heath."

He was being confronted with something that happened twenty years earlier, and which he considered just a tiny, insignificant parenthesis to his much more comprehensive project. Just when he thought things were getting back to some semblance of normality, someone grabs a stick and stirs up a hornets' nest.

The resurgence had come with a 1977 piece in the *New York Times.* "Private Institutions Used in CIA Programs to Control Behavior," the headline read. The article revealed that scientists from various American universities had done contract work for the CIA back in the early 1950s and the two following decades—all to find methods for brainwashing and behavior manipulation. Once it became public knowledge in 1975, the so-called MKUltra program was never far from anyone's lips.

A couple of Senate hearings had already established that the program was launched and sanctioned in 1953 and that, over time, it involved research carried out by at least eighty institutions, including forty-four universities. The CIA funneled money for research through apparently independent foundations, and along the way there were experiments carried out on Americans and Canadians without their knowledge. Some died. When the program was finally closed down in 1973, then-CIA director Richard Helms ordered all the records destroyed. But in 1977, twenty thousand unknown documents appeared in the hands of the *New York Times.*

In its article, the paper had exposed a handful of the professors who had been responsible for the most doubtful experiments—especially, the late psychiatrist Ewen Cameron, who experimented with everything from hypnosis to repeated electroshocks to long-term chemically induced comas on his patients at McGill University. But they had also called Robert Heath, who was a well-known name that appeared in the files. The reporter, Nicholas Horrock, had confronted him with the fact that the newspaper was in possession of documents that named him and Tulane as being involved.

Heath held nothing back, told the story about his meeting with Dr. Gunn, who at that time was the head of the CIA's medical division and who had approached Tulane with an offer after the symposium on pleasure in 1962. Might Dr. Heath consider researching the brain's pain system?

"Disgusting. If I had wanted to be a spy, I would have been a spy. I wanted to be a doctor and practice medicine," Heath had said to Horrock.

That had been a year earlier, but now students were poking around in the same case. They were interested in getting all the details out about the university's relationship with secret services. At first, the dean, Sheldon Hackney, would not comment on the documents but instead issued an official statement. But now the *Hullabaloo* and Nadjari had gotten hold of the papers, and they could see there was something wrong.

Robert Heath stated to the *New York Times* that in 1957 he had tested a drug for the CIA—so-called bulbocapnine—and that the testing involved three Rhesus monkeys. But from the documents, even with all their blacked-out lines, it appeared that there was also a test on a single human being. And now this kid from campus on the other side of the city insisted: "Can you explain this discrepancy, Dr. Heath?"

As if he owed anyone an explanation—he had never done research that required any justification whatsoever. But he took a deep breath and explained in an amicable tone to Douglas Nadjari how things were. When Horrock from the *Times* had asked, he had no recollection of having tested the drug on a volunteer.

"I didn't remember it until I later got my old notes sent to me from the CIA."

At the same time, Heath pointed out to the young man that he had originally signed a confidentiality agreement with the intelligence agency and that, under the Espionage Act of 1948, he was prevented from talking about anything without their permission.

On the other end of the line, Nadjari did not say much but diligently noted this down for the article he would spread across several pages in the *Hullabaloo.*

On his own initiative, Heath explained why he had even agreed to test bulbocapnine.

He stressed how the Cold War was at a high point in the late 1950s, and brainwashing was a huge topic. It was all about identifying drugs or methods that could make people talk, give information—preferably without them knowing that they did it. Then they came to him with intelligence that the Russians were studying bulbocapnine, which is why they assumed it might have potential. They knew already that he, Heath, had tested the drug on cats, and now they wanted him to investigate how it affected the higher mental capacities of human beings—specifically, whether bulbocapnine could impair speech, memory, the sensation of pain, and the will to act. They also wanted to know whether the drug might work more effectively in human beings with "a weak nervous system."

To begin with, he was negatively disposed toward the project. He said very clearly that he did not believe in the drug. But it was no secret formula. It was a drug that was already known in the literature.

"You can't wash brains with bulbocapnine," he had said right out of the gate, but they insisted. He replied that he could do a couple of tests on monkeys, and then on volunteers if it looked safe. But he had not imagined using funding from his own research grants. However, the gentlemen really wanted results before January 1957 and provided a grant so that it could be done. A single Rhesus monkey cost $50 at that point in time, and the animal had to be both housed and fed. The $500 Heath proposed was not that much money, but sweet Mary and Joseph, did it result in a lot of paperwork back and forth. Memo after memo.

Irene rummaged through the records and found the papers, and he could see that first there were injections in the two monkeys and then in a single volunteer: an inmate of Angola Prison. For some years, Louisiana's notorious state prison—boasting a death row, forced labor, and the nickname "The Alcatraz of the South"—was a supplier of volunteers for various medical experiments, including Tulane's psychiatric experiments. Heath had never thought it strange that these men would willingly volunteer. The prison at the end of State Highway 66 was known to be one of the country's worst, and the inmates had always seemed quite content to go on an outing to New Orleans and have a couple of decent meals. On one occasion, they had a couple of prisoners escape. A young psychiatrist was supposed to pick up two inmates and drive them to the city, and naïvely he let them out by the roadside when they asked to take a whizz. Of course they disappeared in a flash.

One prisoner had volunteered for a dose of bulbocapnine, and Heath had no doubt they could get something out of digging into this. At that time in the 1950s, there was nothing strange about getting volunteers from prison, but in the current environment he knew that this did not look good.

Otherwise, as Heath had predicted, the infamous CIA drug was nothing to write home about. Nadjari himself could see from the papers he had apparently gotten hold of that the monkeys in the first round of experiments simply grew sleepy. They were just lolling in their chairs after twenty minutes—only to doze off for the next few hours. Moreover, he could assert that their EEG showed a completely normal sleep pattern.

Then there was the volunteer, the prisoner whom Heath had forgotten about the first time. He was a slender, twenty-one-year-old man, friendly and very cooperative. Heath himself interviewed him while Irene took notes. Of course a film was made of the entire session for documentation. The young inmate took the drug intravenously over a few minutes, while they kept an eye on his pulse. It was normal throughout the entire procedure. After a couple of minutes, he spoke about "a beautiful feeling . . . like being wasted . . . almost like morphine."

When the infusion was complete and he had been given 150 milligrams, he mentioned being dizzy. "Nausea . . . a sort of mild dizziness . . . right now, I don't feel so good."

Heath's contacts got their answer in the form of a detailed report and a film from the laboratory. Then they passed their own internal report along in the system, which Heath had been given along with many other memos and notes.

"Judging from the film, we as laymen deem bulbocapnine to be nothing but a poor replacement for fermented grain mash in liquid form, ingested orally," wrote an intelligence officer to his bosses. In other words, they might just as well pour whiskey down people's throats! Heath laughed out loud—he'd had a sense of humor, that CIA guy.

The law student Nadjari was not an expert in psychiatry—nor was he especially interested in the field. He was worried about the relationship between universities and secret services and their activities on campus. But he put down the phone having been given an impression of a scientist who did what he did for patriotic reasons. It was something the man's country asked him to do. And as he later stressed in his article, the records showed that the various tests "were conducted with the greatest degree of professionalism and care."

Nevertheless, there was something that still didn't seem quite right. There was a whisper around the department that had to do with LSD. It had come to light that Tulane was one of five institutions that, early in the 1950s, received a CIA contract to investigate the drug. The intelligence boys were worried that the Chinese were using it to brainwash prisoners and, by chemical means, thereby succeed in turning good American patriots. It was not unknown that Heath in his day had done experiments with LSD on his monkeys and some of his patients, just as there was something to the rumors of studying the drug's effects on healthy volunteers—who were not exactly difficult to recruit from the students on campus. The question was whether the testing was financed by the CIA but specific documents were lacking. The only thing people knew for sure was that in 1955, Heath's colleague Russell Monroe had a CIA contract with the caption "Clinical studies of neurological and psychiatric changes during the administration of certain drugs." And in 1957 Monroe and Heath published some observations of schizophrenic patients treated with LSD and mescaline.

So, how far into the Tulane group's research did the military really reach? The rumors continued to circulate.

CHAPTER 6

The Secret History of Hedonia

The 1970s were a strange time. My own recollections are pretty innocuous—a child's flickering impressions of school chums dressed in flared corduroys, surrounded by grown men with full beards and Prince Valiant haircuts and women with tie-dyed tent dresses. In retrospect, it seems almost comical, but it was an era that harmed a lot of people. It was a decade of hippies and an era in which the struggle against "The Man" and the visibility and rights of diverse groups emerged with a vengeance. The mood was militantly antiwar and antiestablishment, and any form of authority came under heavy fire. Everyone who was oppressed had to be liberated immediately.

In the United States, the 1960s had been characterized by the civil rights struggle, but in the 1970s the movement expanded from being about the rights of black Americans to other groups—women, homosexuals, and, of course, psychiatric patients. The film adaptation of Ken Kesey's *One Flew over the*

Cuckoo's Nest from 1975 was emblematic of the growing feeling that inhuman things were going on at mental hospitals and that psychiatry was the white-coated enemy. The conflict between the film's protagonist, the antiestablishment McMurphy, and the representative of "the system," the overbearing Nurse Ratched, dramatized a popular cultural conflict.

McMurphy is not insane. He is used to show how the system does not actually treat the inmates, but rather oppresses and debases them. The group of doped-up zombies blossoms as soon as McMurphy arranges an outing from the hospital or gets them booze and sex. But this normalization is slapped down with a strict hand. Ratched and the whole apparatus behind her replies by using electroshock as a punishment—ultimately rendering McMurphy harmless once and for all with a lobotomy. *One Flew over the Cuckoo's Nest* was a success. It won an almost unheard-of number of Oscars in five major categories. It was also a hugely popular success and helped shape an entire culture's attitude toward both electroshock and psychosurgery.

At the same time, interestingly enough, it was in the 1970s that psychiatry as a discipline went through what has been called the great remedicalization. The discipline ever so hesitantly fumbled its way back to its medical and biological roots after having been dominated by the psychoanalytic paradigm since the 1930s. The root of the upheaval was a revolution in medication with the many new drugs that emerged during the mid-1950s and, for the first time, provided resources against psychoses, manias, and depression. The medicine worked so well in spite of everything that it naturally set off a search for the underlying biological mechanisms of these diseases. If a chemical drug works specifically on a symptom, they had to find out what the drug did in the organism and, thus, reveal the cause of the ailment.

Gradually, the documentation that psychiatric conditions were better explained by chemistry than childhood incidents was so massive that it could not be ignored. And the counter-movement spread—particularly from Washington University in Saint Louis and a group of psychiatrists that came to be known as the neo-Kraepelinians because they harkened back to the German founder of modern scientific psychiatry in the 1800s, Emil Kraepelin.

In 1968, Washington University's top psychiatrist, Sam Guzé, published the article "Why Psychiatry Is a Branch of Medicine," and a massive attack on psychoanalysis launched from his department. The group insisted, among other things, that diagnoses must be valid. That is, a given patient must be able to count on getting the same diagnosis from different psychiatrists—something that was definitely not the case. The neo-Kraepelinians worked to establish strict diagnostic criteria and helped shape the first modern version of the field's bible, the *Diagnostic and Statistical Manual for Mental Disorders*, DSM, which was published in 1980. Around that time, it could be said that biology formed the basis for mainstream psychiatry and that psychoanalysis was a more or less marginalized phenomenon.

It is a tragic irony that, during this same period, Robert Heath experienced his own marginalization. Whereas he should have fit in with the predominant tone from the mountaintop, he received seriously negative press for the first time. The first steps toward the blackening of his legacy were being taken.

Today, when you look up Robert Heath on the Internet, you usually encounter references to the idea of mind control, to the CIA, and to the struggle against psychosurgery. There are also a lot of indignant references to the study from 1972, the attempt to convert the young homosexual man they called patient B-19.

I wondered whether Robert Heath's methods were simply too much for his colleagues or whether something else was at stake. How is he to be interpreted?

I discovered that there are a couple of academics who have asked themselves the same question and that one of them has even written a PhD dissertation on Robert Heath. I heard about it by accident from Clarence Mohr, who is a historian, formerly of Tulane, and the author of a book about the university. Heath was mentioned in it a few times, and I wrote Mohr to hear whether he had seen the films that Tulane possesses. Maybe he had an idea how I could get access to them. He responded at once, but he had not seen the films. "Quite honestly, I doubt that I would really want to," he wrote, but he suggested I contact a certain Christina Fradelos, who was a history student some years ago at the University of Chicago. She wrote a dissertation on Heath's electrode treatment: *The Last Desperate Cure: Electrical Brain Stimulation and Its Controversial Beginnings*.

I was able to track down and buy an electronic version of the text and hungrily read the 150 pages or so, eager to know what a professional historian thought.

Her judgment on three decades of experiments behind Tulane's thick walls was not mild. Robert Heath's deep electrodes were simply the last in a series of psychiatry's desperate but poorly conceived attempts to treat the untreatable. The wild ride went from fever therapy and insulin comas to lobotomies and electrical stimulation. Robert Heath was influenced by a desperation inherited from his predecessors in the field, claimed Fradelos, and it made him act irresponsibly with the idea that what was new must, at least, be *better* than what they otherwise had to offer.

"Heath's contemporaries rejected him for repeatedly failing to design experiments according to their scientific and ethical standards," she wrote. And over several chapters, she drew a picture of an ambitious but poor researcher who was inspired by mutually contradictory theories about how the psyche works and who, in a vain hope for a Nobel Prize, set aside any concern for his patients.

The dissertation was from 2008, but today, strangely enough, Christina Fradelos will not talk about it. I pestered her for an interview with a series of e-mails and even cold-called her at her current place of employment, which I dug up from the Internet. My pleas did not fall on welcoming ears. Fradelos had left the academic life and could not spend her work time at a YMCA discussing her old studies. Before she hung up, however, she promised to send me her private number so I could call later. But she never did. A few friendly reminders were met with a persistent silence, and I asked myself how you could spend four years studying something, digging so far down into a matter, and then turn your back on it.

By contrast, another critic of Heath welcomed me. A psychology professor at Louisiana State University in Baton Rouge, Alan

Baumeister settled down to history studies late in his career and published three articles on Robert Heath. He was finished with that topic now, he confided in me, but was glad others wanted to look more closely at the case, which he believed should be brought to light. As he wrote: "The story of Heath has been neglected. For more than ten years, I have tried to get my hands on more documentation but without luck."

Baumeister caught sight of Heath in the summer of 1999—a few months before the old man died. Originally, he wanted to write about how Heath was cheated of the honor of discovering the brain's pleasure center. However, when he buried himself in the scientific works, his focus changed to the ethics of the project. In 2000, Baumeister's account was published under the title *The Tulane Electrical Brain Stimulation Program—A Historical Case Study in Medical Ethics*. Here he argued, like Fradelos, that Heath and his people did not only do bad science, their work was also ethically flawed—as he stressed in the article, "not just by the standards of our day but by the standards of its time."

The Tulane group claimed that it wanted to treat hopelessly mentally ill people who had no other possibilities. At the same time, they did *research* on these people, experiments that were not necessarily for the benefit of the patient but, instead, had to do with the researcher's own curiosity. The ethical precepts for human experimentation that were formulated after the Second World War and the Nuremberg trials stipulated that the research subject had to give his or her voluntary consent. But did they actually do that at Tulane? asked Baumeister. Like Fradelos, he also emphasized that Heath always claimed there was a therapeutic rationale, but that rationale was often very hard to see. Why, for example, place up to twenty-five electrodes in the brain of a schizophrenic when you had scientific support that stimulation only of the septum might have a therapeutic effect?

"Really, I was too easy on Robert Heath," Baumeister told me, calling Heath straight out "a monster." I asked for concrete examples of Heath's abuse, and Baumeister mentioned how Heath did research on patients who stimulated themselves like the rats in the experiments done by Olds and Milner. And even worse: He subjected patients to downright painful electrical stimulation. In the same way I had received a cold shoulder from Tulane, Baumeister was prevented from seeing the infamous films, but he had read descriptions of them—including one by the journalist Judith Hooper, who visited Heath in the 1980s and wrote about him for the periodical *Omni*.

"We're talking about patients who were writhing and shouting that they wanted to kill somebody. That sort of thing has no therapeutic value whatsoever!"

I understood what they were saying, these two academics. But I had a hard time agreeing with their one-sided judgment. I asked myself whether they might not be looking at Heath through today's filters after the development over the interim years of what you could call the ethical landscape. The view of what is permissible in medical research has, after all, changed a lot.

Alan Baumeister himself actually pointed out that none of the otherwise so critical colleagues who gathered in New Orleans in 1952 to hear about the new research had raised a single *ethical* concern. Everything revolved around the science. And even I had a hard time seeing a breach in ethics from the perspective of that time. Yes, the operations were risky, but they took place in an era in which people were still doing lobotomies left and right, a time when everyone believed it was completely appropriate to cut up the frontal lobes of psychiatric patients even though many were injured for life.

Nor was the "treatment" of homosexuality alien to practitioners. At about the same time Heath was doing his experiments, it was not out of the ordinary for families in New Orleans to submit their homosexual sons to electroshock "cures." They received up to forty shock treatments to "erase" the undesirable patterns of behavior, so a healthy and natural sexuality could be constructed. "Regressive ECT," they called it. Other places, they tried to show the "patients" pictures of naked men and, at the same time, give them electric jolts to the testicles.

Electrical stimulation was generally used for things that, today, we would shake our heads at. Alan Baumeister told me—unsolicited—how he, as a psychology student at the end of the 1960s, helped to "train" retarded children to stop harming themselves by giving them electric shocks. I asked, of course, whether at that time and in that context the young Baumeister felt like a monster—the answer was no.

But what about me? This encounter with Heath's critics makes me wonder whether I might be prejudiced in the opposite direction.

Because, honestly, I can't say I'm not attracted to controversy. When I first heard about Robert Heath, the story grabbed me, and it was something that went deep. A haughty contempt for conventions was a fundamental feature of my upbringing. It was a mantra in our home that you should never worry about what "others" thought about you. "Do what you want to do, not what others want you to do," my father might say. Being a yes-man was one of the worst things you could be guilty of, and the people who were admired in my family were those who stood out.

My immediate sympathy still goes to the weirdos, the people who don't just go along but do something *different*. When it is comme il faut to be against genetically modified crops, I have to scrutinize the view and expose the emotional and irrational arguments it is based on. When it is popular to say that there are no biological gender differences, I feel called upon to point out the many and varied differences that research actually indicates. Or worst of all: When the good news is that Danish politicians set the goal that all young people complete an education, I write an article on the most hated research of the day, which revolves around inherited and inherent differences in intelligence.

I don't mind being the bearer of unpopular tidings. What I can't stand is when political considerations or cultural norms are allowed to infect our relationship with knowledge and science, so information is withheld or ignored because it is expedient. The unpleasant things of this world do not disappear because we conceal them but are best dealt with when they have been studied under a searchlight.

Do my own idiosyncrasies, then, influence my view of Robert Heath?

While they got me interested in this story, I keep looking at

Heath not so much as a monster but as a gifted, curious scientist. The people who knew the man and were familiar with his circumstances, these old men, kept me curious.

First it was Frank Ervin and Charles O'Brien, both of whom talked about a man with a humanistic attitude. A man whom they could see had limitations and flaws but whose project and talent they had respect and even a certain admiration for. Later, I tracked down many others and listened to countless hours of recollected memories of youth. The more people I talked to, the more impossible it was to fit their testimony with Christina Fradelos's picture of an incompetent scientist or Alan Baumeister's vision of an unscrupulous monster.

To be interested in a human being you know you will never meet is like reaching out for a ghost.

"If only you could have *met* him," they said as I sat with my tape recorder across from people speaking of Heath as a sort of magnetic presence, a personality that had to be experienced to be understood.

The personal is an overlooked aspect when we talk about the scientific world. We imagine research to be a slowly rolling wave that just washes forward in accordance with its own inner dynamic. More or less anonymous generations of researchers build patiently atop already gathered knowledge and reveal more and more of the hidden reality—almost as if there is some sort of automation to it. But in practice, they are individuals with idiosyncrasies and individual character traits that drive the development and determine its many paths and branches.

Someone decides to go in a particular direction at a particular time. *Someone* has to think differently or break with tradition in order to make some interesting leap.

The entire scientific project hangs to a high degree on personalities. When I think about deep brain stimulation, it is clear that if someone like Helen Mayberg had not been insistent and willing to take a risk—and let her patients take one too, a treatment for depression might not have been developed today and, thus, might not have gotten the high profile and attention it actually has. On the other hand, I've developed a sense that the technology might have ripened much earlier if Robert Heath had had a different personality than he did.

Because who was he?

Words like "complicated" and "complex" recurred with everyone I spoke to—in the sense of an interesting human being with many facets but who was also, at bottom, difficult to understand and difficult to deal with. At the same time, Robert Heath was a figure who loomed large in the lives of his old students and colleagues. Even decades later, they felt a part of his magic circle.

"A day doesn't go by that I don't think of Bob." That was the first thing I heard from psychiatrist James Eaton, who was on Heath's clinical staff in the first half of the 1960s. He kept in contact with his mentor until the end, and wrote his obituary. Today, Eaton has a private practice at his home in one of the better neighborhoods of Washington, DC, but he also has a significant career at the National Institute of Mental Health behind him. While he was there at one point in the 1980s, he wrote to the cream of American psychiatry and asked people to name the five researchers who had meant the most to them. Robert Heath figured in every one of the answers that came back. For some, he had been a positive source of inspiration and, for others, an insufferable provocateur—someone who just *had* to be proven wrong. The point was that no one could remain untouched. As Eaton put it, "Types like Bob push the field forward. It's not the bookkeepers—the people who fill in all the little gaps."

The retired psychiatrist Alan Lipton, who was a student in the early 1950s, stressed Heath's almost monomaniacal focus on his project. Like when he visited Lipton in Miami at the height of the Cuban Missile Crisis, when everyone was talking about an imminent nuclear catastrophe, but Robert Heath simply remarked, "You can't let that sort of thing stand in the way of your research."

And Robert Heath saw himself first and foremost as a *researcher.* The tragedy was that everyone around him could see that he was actually something else. Eaton called him an almost clairvoyant theoretician, who could often draw the contours of connections that were only demonstrated later. Among his proselytes who chose to go into practical psychiatry, all emphasized that Heath was a superb clinician, a therapist who could reach even the most difficult patient and had an immediate understanding of people's problems.

"But as a doctor he was not trained in stringent scientific methods," Colby Dempesy said to me. Irene's brother, who is now eighty-five, was for many years a professor in physics at

Amherst College but worked for some years with Robert Heath in the 1970s. He got to know him as a man who spouted ideas left and right but was bad at differentiating the worthwhile from the rest. He also had a hard time sharing the credit for work with other people and, therefore, was not good at forging collaborations that might have moved him and the field forward. But as Donald Gallant, who was Heath's colleague for many years at Tulane and is now a psychiatrist at the University of Memphis, remarked, "Bob didn't give a fig about what other people thought about things and never accommodated himself to anything."

While he withdrew into himself in his work, he was, by contrast, expansive in the way he lived his life. He was a party animal who sought to make life fun, and was an eternal practical joker. Like when he and Charlie Fontana went into Dean Lapham's office the night before an inspection from the board of health and left a bunch of lab rats they had just fed laxatives. Or when he introduced a Russian colleague at a department meeting. The Russian got up and gave a longish speech with a thick accent about how well known and admired Professor Heath was in the Soviet Union and how they wanted to pursue his research agenda there. Later that day, those assembled realized that the foreign "colleague" was, in fact, Heath's own brother on a visit from Pennsylvania.

I caught a glimpse of the party animal in Alan Lipton's office in Miami, which contained a faded black-and-white snapshot of the two together at a reception many years before. Relaxed and in half profile, Robert Heath was standing with a drink in his hand, his tie loosened, and a smile that lit up the picture but, at the same time, made him seem like a lion checking on his pride. Charisma is what it was; hard to resist. Since childhood, I have been shy; even at the age of fifty, I am still hopelessly awkward in social situations. Receptions and networking events, which seem to be omnipresent and defining for our work relationships, are my nightmare. Groups of people in conversation look to me like impregnable fortresses, and I usually steer myself toward a distant corner, where I wait for somebody to come over to me. Those who do are usually other people who don't quite fit in. Whereas some people cannot step into a room without everyone immediately being aware of it, I am pretty much transparent. So I understand only all too well Alan Lipton's observation that Robert Heath may have had his large personality against him—he may have been hard to resist, but he aroused envy. As he told me: "Bob had *everything*. He was blindingly good-looking, ran a big department with lots of research funds, was incredibly attractive to women, and also a superior sportsman who played fantastic tennis and golf. And on top of that, he had a wonderful family."

I waited a long time to contact the family. I'd been pussyfooting around even though I already knew the names of the five children and the cities they lived in. It was all there in an obituary of Eleanor Wright Heath, Robert Heath's widow, who died in 2012. There was a daughter in Massachusetts and another in Washington, and there was the son, Robert Heath Jr., who lived in Florida. I had tracked down their phone numbers and neatly written them in the margin of the yellowing obituary. Nevertheless, I kept postponing it—which of them should I try first? Maybe it was the name itself that did it, but when I finally got around to dialing, I called Rob Jr.

Stuttering more than I would have liked, I explained my purpose—to get closer to his late father. At first, the other end was quiet, and I hurried to fill the silence by stressing that I was not after any sensational story. It was all about his scientific work.

"Why don't you come out here and visit us? We can meet at Hedonia."

"Hedonia." Written on an old, slightly rusty metal sign on the side of the road outside Picayune, Mississippi, the word suddenly seemed very strange. Beneath it was written "*R. G. Heath*" in sweeping letters, and I almost had a shock when, after I had driven up the country road and into a courtyard, I was met by a tall, slender man with tanned skin and thick salt-and-pepper hair.

The resemblance was unmistakable. Rob Heath Jr. had some of the same distinctive features as his father, and his eyes were intense. But they were also a bit shy. His voice was deep, but the words came out sparingly and cautiously.

"The place was much better maintained before, but the family doesn't come here that often anymore," he apologized as I was still getting out of my rented car. "We are all scattered to the winds."

Rob was a biologist, like me, but whereas my curiosity was piqued by molecules and cells, his was fixed on animals and nature. The rare and vulnerable birdlife of swamps and wetlands captured his heart. Back home in Florida, he worked as a volunteer for the Audubon Society, counting endangered breeding birds.

"There are over two hundred acres of forest land," Rob said, turning slowly as he surveyed the property. He pointed first in one direction and then in another. "Dad had cattle in the back down there, and you had a view directly to the lake over here. Now it's all hidden by trees."

The house Heath had built was made of solid timber, a spacious cabin with a high-ceilinged living room and stairs up to the many bedrooms on the second floor. It had a typical summer-house atmosphere of a certain reserve and a liberating indifference to aesthetics in its decoration. On the walls were trophies from a life lived in the country—an old rifle and painted wooden models of particularly memorable fish that Bob Heath was able to land. Out in the low-ceilinged kitchen where Rob was brewing up some coffee, there were children's drawings and snapshots of the family. On one of them was written # *One Grandpa*, probably by one of the three brown-haired children who surrounded Heath and made him smile in a way I had not seen in other pictures.

Today, the place felt a little abandoned. From Heath's old colleagues, I heard about the tennis court, where their boss excelled with the racket and installed lights, so matches could be played even at night. Now the lamps had been dismantled, there was no net on the court, and thick tufts of grass had forced their way up through cracks in the cement.

"He could have been a professional tennis player. Even when

he was old, I couldn't beat him," said Rob. As we strolled around the overgrown property, I listened to fragments in the story of a father figure who was the center of his life and still a common reference point—a fixed star. A father who was away from home a lot but was intensely present when he was there. A man who handed out nicknames to everyone, who gathered the family for parties at Hedonia, and who loved to tell racy jokes when the old aunts were present.

A father who kept his work strictly separate from his family life but, at the same time, desperately wanted his only son to follow in his footsteps. Again and again, he dragged the boy along to the laboratory and introduced him to his colleagues, the research animals, and especially the patients. The mental patients always greeted the professor's son warmly, and several of them tried to tell the boy how much his father had done for them and how grateful they were.

"I thought the place was dark and sad, and everything was a little creepy," the grown son told me—now from behind his sunglasses. Strange people with electrodes sticking out of the back of their heads, and caged monkeys with sad eyes—if that was research, he didn't want anything to do with it.

"I remember one day back in the 1960s I was sitting alone in Dad's office, fiddling with some test tubes that were on the desk, and 'LSD' was written on one of them. I just put them back and didn't ask anything."

The Heath children realized that Dad was a controversial figure and that a lot of what he did would not sound pretty to the sensitive ears of today. Rob himself remembered how, as a young student at Tulane, he was embarrassed and uncomprehending when the story about the experiment with the homosexual and the prostitute came out. What in the world did that have to do with science?

"But Dad didn't understand the criticism. He reacted by trying to explain why it was a quite rational and excellent experimental treatment."

We had walked some distance from the large house and stopped in front of a freestanding wooden building with double doors and some narrow windows beneath the gutter. It might have been a boathouse, but it was far from the lake.

"Dad's office. He used it when he was working on his monograph."

Rob got a key, and we went in. It was a little like stepping into a tomb. The air was stuffy, and there was a thin layer of black grit over everything. Two strong, cold neon lights beamed down on the 4 × 6–meter rectangular room. One end wall was covered with shelves on which were perched framed diplomas, the sort of thing American doctors were supposed to have a lot of. On the bottommost shelf was a large black-and-white portrait of a round-faced man with a large nose—it was Sandor Rado, Heath's mentor, who in 1955 inscribed a greeting to his "outstanding student." Above the desk in one corner hung a large picture of three tuxedoed men. In the middle was old Dr. Heath—father of Bob and brother Earl—who had his medical practice in Pennsylvania. He was a drinker who died early. Ironically, he looked like W. C. Fields. The lanky boys, who were standing on each side of their father's chair, looked terrified but also hopeful. Beside the photo of the Heath men hung a smaller photo of a very young woman with dreamy eyes and wavy blond hair. This was a professional glamour photo from early in the twentieth century.

"My grandmother. She sang opera before she was married," said Rob.

He walked back and forth, trying to remove a bit of the dust and grime while I looked around. There might be something important in this room, and I didn't want to miss it. I went over to the rows of book spines. Of course Heath's own volumes were there, but also medical textbooks, a little philosophy, and some novels. At the end of one row, like a large bookend, was a black cardboard box marked PATIENTS. I turned around and looked at Rob, who nodded.

Beneath the lid was a row of brown envelopes with black inscriptions—*LSD experiments, monkeys* was on one of them and, on another, *B-14, E. Grant, hallucinations.* Several were mentioned by name and, in the envelopes, were segments of EEG strips and summaries of diagnoses and medical case histories. A stack of human lives, told in brief and boiled down to almost nothing.

"Look at this," said Rob, coming over with some printed pages. "This is from the tribute after my father retired from Tulane."

It *was* a tribute with words of praise from colleagues and thanks from the dean. I had a hard time fitting this warm tone with the wariness I myself felt from the university. I told Rob about my recent visit to Tulane and about how they were afraid of criticism. I had been there to persuade them to let me see the films, which are apparently under lock and key in the department. Hours and hours of footage from over the years. Over an entire afternoon, I'd had audiences with, first, Dan Winstead, the former head of psychiatry, and, then, four of his most trusted staff members. It was an interrogation in five parts in which I was questioned about my project and my intentions. It had to do with access to medical history itself, I argued, and it could only bring illumination and benefit to the debate about deep brain stimulation.

Yes, well . . . was the only reaction I got. Despite their nervousness, I could feel the goodwill from the five academics; but, as Patrick O'Neill, the interim department chief, explained to me, it was ultimately up to the administration and the university's lawyers. "For them, it is a question of image management," he added in farewell. A couple of weeks later, I found O'Neill's signature at the bottom of a letter printed on thick official paper with the decision of the lawyers.

> Dear Dr. Frank. We cannot accede to your request to examine Robert Heath's research material. The patients who participated in these studies did not provide the requisite consent that would allow the university to accede to your request.

Shortly after my visit to Picayune, I received a message that made me laugh out loud at my computer screen. It was from Rob:

> I've gone through my father's office at Hedonia to clean up. I found some old videotapes in a box. A bit of everything. Including footage of patients. Should I transfer them to a DVD and send them to you?

Five brown envelopes arrived, each with a numbered DVD and accompanying notes in neat cursive script. I felt like I was in possession of unopened treasure and didn't know where to begin. Finally, I just took number 1 and put it in my computer. At first, it was just grayish flickers. Then, there he was, standing in the middle of the picture with his hands in the pockets of a white lab coat. It was an old Robert Heath. He had a slight hunch, but his gaze was intense and direct. We had to be in the 1990s, because he explained that what followed was edited footage taken over his entire career to illustrate his monograph *Exploring the Mind-Brain Relationship*. He gave no indication that the contents were anything controversial. There was no apology for the experiments, only for the technical quality—"some of these films, after all, are forty years old."

It was all just as you'd expect with a videotape that has been kept fifteen years in a dank, moldy cardboard box. The pictures were strangely cramped and sometimes flickered out a bit. But the sound was clear. The voice was round and soft, an excellent doctor's voice, calming yet able to issue commands if necessary. The diction was old-fashioned.

First, there were EEG strips. Yards of paper that crawled by at an even pace, while Heath moved a pointer across the curves with running commentary.

"Here, we see an EEG of a patient with acute schizophrenic symptoms of paranoia and confused thinking where an increased activity in the caudal septum is seen. Here, then, comes a reaction with activity in the hippocampus, which is a part of the aversive system and connected to discomfort."

Cut. Suddenly a black sign announces that it is November 1953. *Interview before stimulation*, it reads. I can see from the accompanying notes that it is Joe and that he is receiving stimulation of his tegmentum. He is lying in a bed with a small white bandage around his head, while Robert Heath remains outside the picture.

"Tell me about your weekend, Joe."

Joe had watched a college football game, and they spent a few minutes discussing stats and players. Then there was a sign relating that stimulation at 1 mA was beginning.

"It's like my eyes are hopping," said Joe, who began to breathe heavily. He looked as though a weight was pressing him down.

"Tell me what's happening, Joe."

"I'm seeing double, have a hard time breathing, I feel like . . ."

"Tell me what you're feeling."

"I feel *fierce*," he said, staring suddenly at something outside the picture. "I'll kill you, Dr. Llewellyn. I just want to *kill* someone!"

He whined and nearly howled, raised himself up on the bed and began to rip into the sheets. I realized that this was the footage Alan Baumeister was talking about as a breach of all ethical rules. It *was* difficult to watch.

"I'll rip this shit into a thousand pieces," shouted Joe, who broke out into a furious tantrum. As uncontrollable as a child. But suddenly, he fell back as if something had let go of him. The stimulation was over, and Heath placed a hand on his forearm and began to ask him about the experience. Joe related how his hands felt like claws, and he hadn't really wanted to harm Dr. Llewellyn.

"But whoever was standing there looking at me would have gotten the same reaction. It was like a force that was unleashed in me."

Cut. A new black sign said that it was now an hour later, and Joe was leaning back calmly with one hand behind his head. He looked up thoughtfully at the ceiling and related how the painful stimulation reminded him of the fits of rage he had out in the real world every so often. In fact, he was specifically reminded of an episode in which his sister was supposed to iron a shirt for him but scorched it around the collar.

"I just grabbed it and ripped into little pieces. I had to do it, it was like some uncontrollable force," he said again.

To me, it seemed like an ordinary therapeutic conversation. The psychiatrist who asks his patient to explain what he is feeling and what the feelings make him think. It was quite clear that Joe understood and accepted the treatment and was cooperating. He tried to describe as accurately as possible the "murderous anger" he struggled with and how it felt when he was seized by it. It struck me how everything was reminiscent of modern-day cognitive behavioral therapy. CBT took shape—and took off—in the 1980s. It is now widely used to treat anxiety, mood disorders, personality disorders, and even sometimes psychoses. It is decidedly not about navel-gazing analyses of "deeper" or "hidden" psychological causes but quite simply tries to identify negative or erroneous patterns of thought and change them. Patients learn to recognize when and where things go wrong for them and are taught strategies to steer clear of their personal pitfalls.

Then the film broke and flickered again, so I fast-forwarded in small jumps. Once again, we caught the image of Heath in front of a white wall. The patient we were to see, he explained, suffered from a chronic low mood and was usually angry and irritable. "Notice the change when the stimulation of the septum begins. The patient begins to smile, and his associations become more positive."

Cut to a black sign. *Patient B-19. Temporal lobe epilepsy with paranoid psychotic behavior.* It was B-19! The famous homosexual.

He was young. His hair had been shaved off for his white cap, but he still had a bushy, dark mustache. He was lying on his back in a white bed, dressed in a white hospital gown, his hands folded over the blanket. He was staring up at the ceiling. It was strange to see him in real life. I didn't know what I had expected, but probably a confused and desolate person. This guy seemed calm and at ease with the situation and totally without resistance. He was reflective and spoke as though he was engaged with the psychiatrist, who was outside the picture.

"I wanted to tell you something that recurs in my life. Every time I have encountered something good, something I could look forward to and be happy about, it was always overshadowed by something unpleasant."

"Try to describe that more."

"Like when I was six years old and got an electric train for Christmas. It was *exactly* the train I had wanted and looked forward to, but I couldn't enjoy it. There was just something missing."

He turned his face toward Heath and shook his head,

reflecting on himself. Then he lay back on the pillow, and a smile erupted in the middle of his melancholy recollection.

"What are you smiling at?"

"I don't really know." But his smile broadened, and when Heath asked again, he responded with a sort of giggle.

"It feels like when I smoke pot, and I get silly and think of something funny."

"Pot, you say?"

"Yes, but right now I'm actually thinking about my electric train. How I saw it in a shop window and just wanted to have it."

"Good to see you smile."

"I'm better than before. I'm more interested in things, I think. It's like a little island of happiness has appeared. Charlie is doing something out in back, isn't he?"

"Why do you think that?"

"Hah. Of course, you wouldn't say anything. But I can feel the change in my mood."

Cut. Black screen. It seemed as if there was no more, but finally the picture appeared again. It was an EEG strip with its spindly ink tracks from twelve electrodes, and Heath's voice broke through in a sentence: ". . . recording from the brain during actual intercourse," he said.

"In order to illustrate the changes that are taking place with the intense pleasure that an orgasm represents."

My God. It had to be the recording of the notorious experiment with the prostitute. The stream of paper crawled slowly forward, and Heath used his pointer.

"Here, we see so-called spindle activity in the septum."

The pointer follows some tall, wide waves.

"The patient is in a pleasant mood and is looking forward to the meeting with a . . . er, um, partner.

"Now he is moving somewhat. This is seen by the activity in the electrodes placed on the scalp. He is restless. Now he is discussing with his partner about their coming, er, relations, and we see this response from the amygdala . . . here."

Some small, aggressive-looking spikes replace the flatter tracks.

"Now the patient tells his partner how good it is and says 'oh' and 'ah,' and there is a significant response in the septum . . .

"Now he is complaining that he feels a bit weird . . . then there is movement again . . . some foreplay is taking place at this point, which we can hear on the tape. 'It feels good, really good,' says the patient."

The pointer made a click against the paper. "Now we can see steadily increasing activity here in the septum and the dorsolateral amygdala . . . it is building up to an orgasmic response. Here—now we see a rhythmic amygdala activity that is followed precisely by the activity in the septum and the orgasmic culmination."

The almost flat line was transformed into a dramatic mountain landscape.

"Now he begins to move and is quite satisfied. 'Do you mind if I shout out the window?' he says. It is the first time in his life he has achieved an orgasm with a woman. He keeps talking about how happy he is and also receives compliments from his partner . . . they have a pleasant conversation . . . it continues here . . . and here we see again normal EEG activity, which continues."

The pointer disappeared from the picture, the strip kept running, but I stopped the film. It felt terribly intimate. I had just followed the fine ink line like it was Morse code sent out from the innermost part of another human being. I had not seen the young man with the mustache in the physical act but had stared straight into his raw emotions as they unfolded.

I quite understood the urge to condemn—this *was* really perverse, you can't help but think. But why?

We do not see sexology as something reprehensible—not even the sexology of the past. Just think about William Masters and Virginia Johnson from Washington University, who were contemporaries of Robert Heath and are currently celebrated as pioneers in sex research. Many presumably know them from the popular and critically praised HBO series *Masters of Sex*, where they appear as a couple of genuine heroes for sexual freedom. The series shows how the two defied the straightlaced university administration and insisted on illuminating the mechanics of human sexual behavior. This included hiring Saint Louis prostitutes and visiting brothels in order to "learn from the experts."

Masters and Johnson worked between 1957 and 1990 to uncover details about the physiology of the female orgasm, and their laboratory data was based on "10,000 cycles of sexual response," as they put it—that is, masturbation sessions of 312 men and 382 women, conducted with various measuring devices while the researchers watched from behind a window. In keeping with their time—but omitted from the current TV series, Masters and Johnson had also conducted a therapy program to convert homosexuals, a task they even claimed to have had success with in 7 out of 10 cases.

In 1966—six years before Robert Heath's experiment with B-19, the two pioneers published for the first time their collected results in the book *The Human Sexual Response*, and it was a bestseller. Today's researchers are still interested in the anatomy of sexual enjoyment. Author Mary Roach investigated modern sex research in her book *Bonk* and, among other things, she took her husband, Ed, into an MRI scanner, where the couple had intercourse in the narrow tube. In the name of science and knowledge, of course. And on YouTube, you can see a film of a copulating couple, created by the Dutch researcher Pek van Andel, who was quite simply interested in understanding the actual dynamics of the sex organs during the act. The film brought him an international fan base as well as a much-sought-after Ig Nobel Prize, which is awarded to unusual and spectacular research with comic implications.

It made me think of a conversation I had with Charlie Fontana. We were sitting in my hotel room with a view of New Orleans, and the old technician told me that he'd had a great deal of contact with homosexual men in the 1970s. Robert Heath had them in psychiatric treatment, and they were sometimes there in groups.

"I don't recall that any of them *wanted* to be homosexual," Fontana said. He also said that he could clearly remember B-19 as "an extremely interesting young man." He stayed at Charity

Hospital for months and was always committed to "the project." Fontana kept talking about how it was Jim Garrison, New Orleans's prominent district attorney at that time, who granted permission for the experiment with the prostitute. Garrison (who was quite bipolar and, at one point, himself in treatment with Robert Heath) even helped to find the young woman who participated. Suddenly, the eighty-three-year-old man sighed, and a tired expression appeared on his face. Then, he said in reference to his boss: "Maybe he shouldn't have done that experiment."

You should learn from the past, said Helen Mayberg, and there I was, all alone with a historic treasure that very few had seen. I looked over the stack with hours of footage, wondering how much potential learning was hidden there, when my gaze was caught by something unexpected. It was a DVD, and written on the back was *Cerebellum*. The word is the Latin designation for the "minor brain," the curly appendage that is squeezed between the back of the brain and the medulla oblongata. I wondered what in the world the cerebellum could be doing in this story.

CHAPTER 7

A Cure for Violence

Helen and John Merrick were on tenterhooks, each sitting in their own armchair in Robert Heath's office. The middle-aged couple was in despair. As good Catholics, they had prayed and prayed for their son David and, as good parents, they had done what they could to raise him and cope with him, but nothing helped. The boy was beyond reach, destroying himself and the whole family. They were ready to take drastic steps—even steps that might make many of their religious friends cross themselves.

"But we have thought it through," said John. "If this is the only chance our son has, let's take it. If we can give the boy a possibility of living at least a *partially* normal life, it's worth it."

Helen took up their plea. With a quavering voice, she told Heath that the most important thing for them was to do something about the *violence*. David was harming himself. He actually tried to kill himself several times, and she was always afraid he would hurt other people.

"I . . . I am even willing to risk his life if an operation can do something for him."

Robert Heath nodded understandingly and explained what he could offer. It was 1976. He had just gone through the media grinding machine, and this made him sensitive to ethical requirements—such as informed consent. On this day, he even filmed the conversation with his patient's parents in order to document precisely how they made their decision on a qualified basis.

David was going to get a new device. A brain pacemaker that Heath would test for the first time. It was a chiplike plate with ten platinum electrodes. It would be implanted on the surface of the cerebellum. "About right here in the neck." He showed them with the flat of his hand. From there, it could send short pulses of weak electric current into the boy's brain and, hopefully, stop the worst of the fits.

David Merrick was only seventeen years old but, inside the system, he had been designated as "the most dangerous psychiatric patient in the State of Louisiana," and the stories about "that evil boy" flourished among the hospital personnel. Even though he was small and slender, he could go berserk, and it might take eight caregivers to hold him down, they said. Stories were told about the time the police arrived at the Merrick home in Mandeville to take him to the hospital. The boy tore the door off their patrol car before they overpowered him.

David had bounced from institution to institution ever since he was thirteen, and he had reached his end station—the state hospital in Jackson. There, they medicated him until he could no longer stand up. For safety's sake, they kept his ankles chained to a post in the floor. He had spent a year that way when a young social worker overheard some doctors saying that the boy would not survive another year under those conditions. That spurred her to contact the Merricks with a message to visit Robert Heath at Tulane. If anyone could do anything, she said, it was him.

David had never received any real diagnosis. But he had turned blue due to lack of oxygen shortly after birth, and his mother noticed that something was wrong with the boy ever since she brought him home from the maternity ward. He screamed and screamed in a way that her three older sons had never done. For the new mother, it sounded as if the boy was being tortured. But no matter what she tried he continued to howl almost like an animal until he fell asleep on his own. Night after night.

When he got a little sister two years later, it was clear that his development was following its own trajectory. While the new baby learned to speak normally, David remained silent and only began to utter a few coherent words when he was five years old.

On the other hand, he was physically out of control. He wrecked his toys and the children's furniture. If he had the chance, he would set fire to things. One moment, he would be playing peacefully with his sister, and the next he would be banging her head against the wall until a grown-up stopped him.

His parents consulted doctor after doctor, but they received no answer except that David was a no-good kid—a "bad boy," right to the bone. But there was something mysterious and, at the same time, heartrending for the family about the boy's violence. If he was not caught up in an outburst, he was a gentle boy who cared about his sister and was very good with cats. David had no explanation for his conduct but just repeated to his mother that he did not do it on purpose. He didn't *want* to be bad, but it was like "something" inside him made him do it. Something he couldn't describe that took control of him.

Can you exorcise violent demons and strangle uncontrollable impulses with an electronic chip in your neck?

Robert Heath was convinced that it must be possible, because he had made a discovery. He identified some hitherto unknown connections in the brain and found that there was communication between the outermost layer of the cerebellum and the emotional areas far inside the limbic system. Precisely the areas he had focused on from the beginning.

The idea itself went against everything the textbooks dictated. Traditionally, the cerebellum was considered to be merely a sort of appendage taking care of such bodily tasks as keeping a check on movement and balance. This little structure was supposed to coordinate countless sensory-motor inputs from the whole body and regulate the force, rhythm, and accuracy of our movements.

But now Heath opened up the possibility of finding a direct line between bodily sensations and the emotions. He was certain that he could influence the central emotional areas—the septum, hippocampus, and amygdala—by stimulating the easily accessible cerebellum. This was far simpler and safer than trying to place electrodes deep inside the brain, and he felt the pull of new and great opportunities. This was just the ticket for achieving results and reaching patients with whom he had not yet been successful, and he could imagine the recognition that should—that *must*—come with it.

The line of research began in 1972 with a phone call from an old friend and colleague. The man was a psychologist at the National Institute of Child Health and Human Development in Washington, DC, who was interested in understanding how and why children developed severe emotional problems. He had gotten hold of some monkeys that he believed might be the key to some answers if only Bob would help him. They were five Rhesus monkeys of the same species that Heath already had at his facilities, but these little guys were different. They were completely screwed up in the head. They lacked any normal social behavior and conducted themselves almost as if they were autistic or even psychotic. Couldn't Bob find a way of putting electrodes into their small brains and figure out what had gone wrong?

The friend from Washington knew perfectly well that he would grab Bob's attention with the word "psychotic," and he was right. Robert Heath could see how he would be able to record the brain activity from the disturbed animals and compare it directly with his own normal Rhesus monkeys, who carried around permanent electrodes. He had just one question: "Are they Harlow's monkeys?"

They were. These socially maladjusted animals were sought-after creatures. They were the result of some famous—and eventually infamous—experiments conceived by the psychologist

Harry Harlow from the University of Wisconsin–Madison. For twenty years, Harlow had been preoccupied with mapping the "nature of love," as he called it. He investigated how circumstances in early life shape an individual's later psychology. He had used drastic means.

In the early 1960s, Harlow and his wife, Margaret, had done experiments with total social isolation in which they removed newborn Rhesus monkeys from their mothers and, for periods between three months and two years, left them utterly alone in cages without any form of toy or object. After six months, the animals had already suffered irreparable damage. And the longer the isolation lasted, the more abnormally they developed.

The disruption of their behavior began almost at once. The tiny creatures developed so-called stereotypic behavior and walked like robots back and forth in their cages. Or they sat all day long in a corner, rocking back and forth in the way autistic children at the institutions of that time were known to do. When the monkeys were finally taken out of isolation and put back into a group, everything went wrong.

The isolated animals had never learned anything about social signals and could not get along with fellow members of their species. A few were so overwhelmed that they refused to feed themselves and died in a few days—of "emotional anorexia," as Harlow noted. The general pattern was that the psychologically broken animals were in a chronic state of alarm and terrified of any touch or approach from others. Many exhibited self-harming behavior, and some had a tendency suddenly to explode in aggression. Without any visible occasion or provocation, they might attack random individuals or things furiously. Even something as fundamental as mating behavior was destroyed. The isolated males never succeeded in mating because they were incapable of and apparently uninterested in attracting a female. The researchers could force the females into mating by binding them to what Harlow dubbed a "rape rack." But when their offspring arrived, they were indifferent to them. The isolated females ignored their young or mistreated them.

These studies and their glaring results had an impact in several ways. Among the public, the methods themselves provoked outrage. Arguably it was the starting gun for the animal welfare movement. In the academic world, however, people mostly looked at the results, and they were particularly appreciated among psychoanalysts. Finally, *finally*, Freud's disciples could point to some apparent evidence for the classic theory that psychotic behavior was the mother's fault, the idea that the cause of schizophrenia was to be sought in an upbringing with a cold and distant "refrigerator mother."

But the enthusiasm didn't last long. Harlow did later experiments in which he allowed young monkeys to grow up without a mother but in each other's company—and, under these conditions, their psychological and behavioral problems were actually quite subtle. This indicated that it wasn't Mom and her cold ways but social intercourse generally that was fundamental for normal psychological development.

But still, not entirely. Harlow's colleague William Mason discovered that there was something more going on than just sociality. He had a feeling that access to sensory stimulation also played a role, and late in the 1960s he did an experiment aiming to clear this up. He compared Rhesus monkeys that had grown up in three different sets of circumstances. The babies stayed with their mothers or they were transported after birth to a cage with a mechanical surrogate mother in the form of a large plastic bottle. This "foster mother" was either wrapped in nice, soft fur, standing still in a fixed place, or she had no fur but was motorized and in constant movement.

Something interesting happened. The monkeys clung to their "mothers" in both cases, but they got something very different from it. The ones who grew up with the nonmoving mother developed pretty much like Harlow's classic isolated monkeys and ended up deeply disturbed. The babies with the moving surrogate, on the other hand, developed far fewer and less striking disturbances. Mason had to deduce that *movement* in early childhood was crucial in some way for normal psychological development. The question was why and, in particular, how?

Robert Heath saw a heretical connection to the cerebellum. It seemed an obvious thing to assume that this movement-oriented structure developed abnormally if it did not get sufficient motion stimulation. Might it not have to do with the fact that an abnormal cerebellum gives rise to abnormal psychology because the cerebellum was, in fact, connected to the limbic system and thus influenced emotional life?

This was actually an idea he had entertained since the early days. He had discussed it with colleagues. But he had never had anatomical justification.

"Think about little kids," he said to Charlie Fontana. "They love hanging upside down and can't get enough of romping and being swung around—the stimulation presumably goes directly to their pleasure centers!"

He waited impatiently for the five monkeys that would provide them with some answers. As soon as they arrived from Washington and had been accommodated at Tulane, work began. The animals were allowed to habituate themselves to the new conditions—as much as they could, given their temperament—and, then, had a small forest of electrodes implanted. There were all the usual ones—for example, connected to the septum, hippocampus, and amygdala; then there were new electrodes connected to several layers of the cerebellum.

The first measurements were already clear. Surprisingly clear. Heath looked at Fontana over the head of a Harlow monkey that was buckled into its little chair.

"This one looks just like our schizophrenic patients when they are in the middle of a psychotic episode," he whispered excitedly. The disturbed monkeys showed the same powerful activity in the septum and in the parts of the amygdala and hippocampus that were connected with painful and unpleasant experiences in patients. They also had clearly abnormal activity in their cerebellum. The EEG strip came out like a rugged mountain landscape.

Everything looked just like he had hoped. Now the great map-making work began with monkeys, cats and, eventually, rats brought into service. Robert Heath wanted to know exactly how the different areas of the brain were connected in there, in the dark, and thus how they influenced one another. He stimulated all corners of the cerebellum and measured the effect throughout the brain, tracing the path of the signals through the tissue. And look what happened! The activity spread with very short delay to areas in the limbic system. So quickly, in fact, that there had to be a very direct connection—perhaps just one synapse—between areas in the cerebellum and emotional regions.

But the most interesting thing was the effect that emerged. When he sent current into the little wormlike part of the cerebellum called the vermis, he got two opposite effects at the same time. He *stimulated* the activity in the septum, as it was characteristic for pleasure, and he *inhibited* parts of the hippocampus and amygdala, which were connected with unpleasant feelings. Not only that, he could effectively stop epileptic activity that spread out from the hippocampus by stimulating the cerebellum. It looked like an ideal treatment for deep emotional problems.

Everything he had been working on for all these years seemed to click into place, and there were days when he felt ecstatic about his work. Robert Heath was in the autumn of his career with retirement on the horizon, but he felt young again. He was his 1950s self again, back when he first had the burning vision of connecting a healing electrical current to the human brain.

Right in the middle of the intoxication of these new realizations, the Merricks arrived with their ungovernable son, like a gift from heaven. The more details Heath heard, the more certain he became in his hunch that a breakthrough was lurking here. David was quite simply the perfect candidate for the new procedure.

The boy was brought in to Charity Hospital, where Heath put him through a detoxification program, and he began his own investigations. From his clinical observations and yards of EEG measurements, he ascertained that, beyond some slight brain damage—perhaps due to lack of oxygen during birth—and retardation, the boy suffered from severe psychomotor epilepsy.

It was the epilepsy that was giving him the almost psychotic fits of rage, and all Heath's tests on monkeys and cats indicated that they ought to be squelched by applying a rhythmic current to the boy's cerebellum. Still, it was a risky enterprise. There were no precedents to go by, and Heath was already struggling with a local reputation for doing bizarre experiments. A couple of his closest colleagues advised him strongly to forget all about it, but the Merricks pressured him with apprehensive letters, begging him to help them, and his conviction was too strong to let the opportunity pass.

Heath decided to take the leap.

Together with Charlie, he rigged up a pacemaker. Now over twenty years since the two men began to work together, they no longer needed to make the devices in their own machine shop. They could just order the parts from a company, like an assembly set. There was a little plate with the electrodes and accompanying antennae; there was a receiver, which was placed beneath the skin on the chest; and there was the stimulator itself with a battery the size of a cigarette pack, which had to be carried outside the body.

• • •

The planned operation only came about by the skin of its teeth. Not because there was professional resistance but because some local Jehovah's Witnesses got wind of the project and tried to stop it. It was diabolical to penetrate into a human being's most sacred, innermost parts with cold electronics. The organization went to the courts and summoned John Merrick as the responsible party. They wanted an injunction. On the same morning the surgical suite at Charity Hospital was booked, the nervous father had to appear in a court building in downtown New Orleans to defend what he had consented to.

A few hours later, after the arguments for both sides had been provided to the judge, Merrick called Tulane to say that the court had dismissed the case. The message immediately went to the surgeon, who grabbed his scalpel and got going. The operation took longer than he had planned because it was a new procedure. However, by day two, the patient was up and walking a bit. His parents clapped their hands, and the mother spoke in a high pitch about a "miracle of the Lord" while Robert Heath and Charlie Fontana looked at each other in silence. They knew it would be eleven days before they turned on the pacemaker to see whether a miracle had actually been achieved.

In the interim, it was clear that David was still in torment—he was just lying in bed, doing nothing, but he was sullen and uncommunicative. When he could get away with it, he would harm himself or things around him. His body would shake and twist and writhe involuntarily with dyskinesia due to the years of having been subjected to huge doses of antipsychotic medication.

But it all gradually faded when the pacemaker was allowed to transmit its regular signal. The dyskinesia disappeared last, but it was as if David immediately perked up. Over the course of just a few days, his constantly smoldering rage had given way to chattiness.

David could now live at home with no problems, and there was no talk of an institution. Psychological tests and IQ measurements demonstrated that, some months after the operation, he showed a considerable mental upgrade, and for the first time he was enrolled in vocational training. The boy wanted to learn gardening, and everything looked bright.

Right until he started getting into fights with the other students, throwing gardening tools at them and generally behaving impossibly. Again. When he came home, he shouted at a waving neighbor that he should "find something else to stare at" and otherwise go to hell. One evening at dinner, he jumped up, grabbed his mother, and threw her to the floor, where he hit her in the face and kicked her in the stomach before his father could drag him away.

The parents feared that the operation had been in vain, and their family doctor was quick to assure them that it was really just what you could expect with such a crazy experiment. David was and would remain a "bad boy." But the studies at Tulane showed something else: The wires connecting the electrodes to the receiver had broken; the pacemaker was getting no juice. They had it repaired, and the violence immediately vanished.

This way, David Merrick provided his own control experiment. When his pacemaker functioned, he functioned. When it was turned off, he went amok. You could hardly imagine more elegant proof that it really *was* the current to his cerebellum and not any placebo effect or hypnotic suggestion that did it. Robert Heath had a star patient. Here in his declining days and in the middle of increasing problems with the changing zeitgeist, he finally had a *case* he could put on display. Both he and the Mericks got media attention. Not just the local papers but also the *Los Angeles Times* wrote about the new brain pacemaker and its potential.

The stories and word of mouth brought new clients to the shop. Over the course of the next few years, Heath implanted his pacemaker in another ten patients with "severe behavioral or emotional disturbances." These were people who showed up willingly. Typically, they were referred by desperate doctors who could no longer do anything for them, or they came because people close to them had heard rumors or read the articles. It was a mixed bag. There were three whose problems more or less resembled David Merrick's, and there were two women with severe depression that neither medication nor electroshock could help. Finally, there were five schizophrenics.

With them, Robert Heath returned to his core group, the people in which he had the most intense interest. He'd had plans for them, back when it was all still based on a few hopeful studies of psychotic monkeys. Take, for example, the young man referred to him from a psychiatric colleague at the University of Alabama. He was a difficult and tragic case. He had a degree in physics and was involved for a time in NASA's space projects, but delusions and anxiety had kept him home for years and made him incapable of doing anything at all. Trapped inside the four walls of his home, he had attacked his wife several times because his inner voices demanded it, and antipsychotic drugs did not help him.

The pacemaker worked. It removed both the voices and the anxiety and made it possible for the man to throw away his hated neuroleptics and all their side effects. To his own surprise, he was even struck by the ability to feel joy and outright anticipation. He began to speak about finally finishing the PhD he never managed to complete, and he spoke out loud about his dream of a Nobel Prize in physics sometime in the future.

Nevertheless, in 1977, Tulane had to call a press conference. It was not just Jehovah's Witnesses grumbling. A local journalist had written about Heath's latest project as "Nazi research." One warm afternoon in June, Heath had to appear with the Merricks to explain himself. In particular, it was about putting some clear distance between his work and the psychosurgery that got people up in arms at that time. He carefully distanced himself from lobotomies. As opposed to the old operation, he explained, his brain pacemaker did not render his patients emotionally handicapped but preserved their entire register of feelings intact. What he wanted from his treatment was to free his patients by removing the inhibitions their illness imposed on them. And making a direct reference to the obsession at that time with mind control, he made assurances that the pacemaker "would not be able to overrule a human being's free will."

Resistance continued. The Merrick family was approached repeatedly by people wanting to convince them that their son had been a guinea pig for a madman. Word of the harassment reached Robert Heath and he also heard that younger colleagues at Tulane were badmouthing him. But he did as he always did and pushed the criticism to one side. So, despite the fact that there was much less research money than in the good old days, his otherwise ailing program for brain stimulation was undergoing a renaissance in the late 1970s. Between 1976 and 1979, a total of thirty-eight patients had a pacemaker placed on their cerebellum. Along the way, Heath developed an improved version that no longer needed antennae but had a discreet battery inserted under the skin, and it only had to be changed every five years.

For years, he followed patients who had depression or psychoses as a result of schizophrenia, brain damage, or epilepsy with violent impulses. He presented his results at regular intervals at science conferences and published them several times in the well-regarded journal *Biological Psychiatry*. In the last report more than half the patients showed significant improvement of their condition, while the rest had a small or no effect. Once again, he had to admit that it was difficult to deal with schizophrenics—most achieved a certain effect, but many chose after a time to shut off their stimulator. They were too sick and too confused. But the chronically depressive, who struggled with anhedonia—for them, there was steady success—and the same held true for patients with violent behavior.

There had been problems with the equipment itself, Heath admitted, but they had provided something important: The patients had been their own control experiments—controls that he could not ethically put in place on purpose.

This time, there *had* to be a reaction. A response.

CHAPTER 8

Dreams from DARPA

"Miss Lone? Hello and welcome!"

I had only seen David Merrick in forty-year-old film clips showcasing his violence, but I recognized him at once. His hair had grown thin; it was combed over. He was balancing a large pair of brown glasses on his nose, but his round, slightly childlike face and his cautious smile were the same. His light voice was also intact. Back home, I had observed years of David's life as a boy as they flew by in black and white, and now I was in a nursing home in Harvey, outside New Orleans, extending my hand to a sixty-year-old man in a screaming-yellow T-shirt. He took my hand but released it abruptly when he saw the older gentleman behind me.

"There is Dr. Richardson!"

Donald Richardson had driven me over from the other side of the Mississippi River because he wanted to see his former patient. After all, it had been years since he last checked in on the pacemaker that had been put in place forty years ago. But before anything else, David took us on a tour of the linoleum-tiled hallways of the institution—past silent, robed residents to an out-of-the-way corner with three chairs and a TV. There was a sports program on, but Richardson pulled the plug so it would be quiet.

"This is a pretty good place to live, and the staff is nice," said David. He abruptly began to talk about the past. The hospital in Jackson, where the caretakers beat him, and the detox at Charity Hospital, where it took a month before the sedatives were flushed out of his body.

"I often think of Dr. Heath. And of Herb and Charlie," he said, making the treatment back then sound like a holiday camp. The two technicians let him help record the EEG of other patients, and once he even got to operate on one of the lab research cats. David also received visits from other potential patients who had been offered a brain pacemaker and wanted to hear what it was like to have one. I sensed that this bygone time was a benchmark in David's life, a period in which he was a star and got attention from all sides.

But there was trouble in paradise. David told about the time he had been asked by a local TV station to speak against Robert Heath's critics. There were some religious people—Jehovah's Witnesses, he recalled—who were outraged and made "a helluva fuss."

"Do you remember that, Dr. Richardson? How they said Dr. Heath made us patients into marionettes? Puppets he could control? It was crazy! But I talked with those TV people and said it was a lie. Didn't I, Dr. Richardson?"

The taciturn Richardson nodded. During David's monologue, Richardson sat fiddling with a small transistor radio borrowed from the nursing home's personnel. He then walked behind his former patient and moved the apparatus back and forth over his neck. Everyone was quiet while the radio emitted a regular crackle. After almost a minute, there was a loud blip, and Richardson withdrew the apparatus.

"Listen, it seems to be still working," Richardson said, putting a hand on David's shoulder. David seemed as if he wanted to say something but was a little insecure that I was there listening. I took my cell phone out of my pocket and pretended to be reading messages but could not avoid hearing David's loud whisper. "Um, Dr. Richardson . . . I sometimes feel a little strange inside. As if there is something bad that wants to come out. I think there is something wrong with my pacemaker."

The doctor responded soothingly that it was probably nothing to worry about. But he would talk to the manager and arrange for a proper check.

"David was right," said Richardson, when we met three weeks later. Just a few days after our outing to Harvey, he received a call from David's sister Barbara, who was desperate. David had attacked one of his fellow patients at the home and tried to choke the man, who had to be saved by the staff, who had come running as soon as they heard the racket. Several of them had to hold David down as he yelled and screamed that he wanted to kill someone. The police and an ambulance were called. David Merrick was sedated and hospitalized under watch. Richardson had to explain that it was probably because the brain pacemaker had run out of juice. He had arranged with the home for David to visit him and have his battery changed.

"Just like in the old days."

At eighty-three, Donald Richardson looked fragile with his slender build and grayish-white hair. But he was still working as a neurosurgeon and could still stand in the operating room for hours on end. After many years at the larger hospitals in New Orleans, his work was now confined to a private clinic on the north side of Lake Pontchartrain. Patients came there from distant parts of the United States because Richardson had a reputation, not just for being good but for being untraditional and willing to experiment.

I found him through an experiment. In a 2010 paper in the *Journal of Neurosurgery*, Richardson described how he implanted a brain stimulator in a very young woman suffering from what was noted in the psychiatric manual as "intermittent explosive disorder." Funnily enough, it was a diagnosis that his old colleague Frank Ervin described and named in the early 1960s but also a diagnosis that, since the controversies of the 1970s, it had been difficult for anyone to imagine trying to treat with electrodes.

I asked Richardson how he could attempt such an intervention and publish it for his colleagues without setting off a riot. He responded with a slow, rolling, ringing laughter.

"Hah! Most neurosurgeons don't like to think. Unfortunately, they also don't much care about innovation and trying out something new."

Richardson looked a bit wizened, but he had a caustic wit,

and you didn't have to be with him very long to discover that he says exactly what he thinks and could not care less for any form of political correctness. In fact, I could imagine him working well with Robert Heath.

"Bob was a thinker, and that was very attractive to some of us," Richardson said at one point. "People believe he stimulated patients at random with his electrodes, but he would sit for weeks poring over brain atlases and scientific literature before he did anything at all experimental. And when it finally got under way, he gave us surgeons very precise coordinates and instructions."

Richardson was incorporated into the Tulane group already as a medical student, and he helped operate on some of the first electrode patients. He remembered in particular how it affected him to see a terminal cancer patient—a woman in great pain—become pain-free and quite alert and cheerful after a bit of stimulation in the septum. Young Richardson was captivated.

Later, as an accomplished surgeon, he himself worked with deep electrodes in the 1970s to treat chronic pain, and he discovered, together with a colleague, that stimulation of particular areas of the brain unleashed massive amounts of the body's own analgesics. Richardson was also one of the first to use electrodes on OCD and Tourette's syndrome, and he was often called a loner and a maverick.

"Today, I'm interested in violence and aggression," he said. It began ten years earlier when he was flipping through the journal *Science* and happened upon a paper arguing that violence did not have so much to do with fluffy psychic or social problems but, rather, was a question of brain function. Richardson remembered David Merrick and went searching through the literature. He read everything having to do with aggression and the brain and noticed that all the violent cases that had been studied and described had something in common—namely, very low activity in their orbitofrontal cortex—that is, the part of the cerebral cortex that is right over the eyes.

The British neuropsychologist Adrian Raine had observed the same thing and wrote about it in *The Anatomy of Violence: The Biological Roots of Crime* from 2013. Among other things, Raine compared the brain function in different types of murderers: on one hand, those who planned their crime with precision and, on the other, those who just spun out of control and simply went off.

"And what did it show?" Richardson asked rhetorically.

Well, I answered, the spontaneous perpetrators who could not control themselves had diminished function in the right orbital cortex. That was *not* the case for the controlled criminals who had premeditated their violent act.

"Exactly, they're just psychopaths," Richardson maintained, and he began to rummage around for something on his computer. While I waited, the contrast between the past and present struck me once again. Whereas decades ago Frank Ervin was vilified for arguing that aggression was partially founded in biology in *Violence and the Brain*, that idea is fully accepted now. When Adrian Raine called his book "a manifesto for neurocriminology," reviewers reacted with the word "refreshing." In reality, this is not because the basic data has gotten so much better but because the zeitgeist has changed. The three pounds of quivering tissue between our ears is no longer sacrosanct but merely another component of our bodily machinery.

"Here," Richardson finally said, showing me the circuit of violence he had drawn. Three structures were accentuated: the orbital cortex; the deeper-lying thalamus; and the little almond-shaped amygdala, all of which were connected. The amygdala registered moments of danger and discomfort and communicated its concerns to the thalamus, which functioned as a kind of motor that chose to react with flight or fight. The third wheel was the orbital cortex. It had the task of considering and modulating the urges from the thalamus so they fit with the actual circumstances.

"The left orbital cortex works like an accelerator, while the right is the brake," Richardson said, going back to his own violent patient.

Nikky, as she was called, is now twenty-two, and she sounded to my ears like an updated version of David Merrick. Since childhood, she had regular fits of uncontrollable rage. As she grew bigger and stronger, the problem became unmanageable. When, as a teenager, she almost killed her grandmother with a knife, a long period of hospitalizations began at various psychiatric institutions. There were sedatives, there were physical restraints, and there was a prospect of life as a patient in a closed ward.

That was when Nikky's grandmother turned to Donald Richardson, about whom she had heard. He had her grandchild's brain scanned, and the scanning revealed some slight brain damage and diminished function in the right orbital cortex. Richardson got the necessary permits to implant a single electrode to stimulate communication from the injured area, and the operation went smoothly. For a year, Nikky's condition oscillated back and forth until Richardson found the right parameters. Then she calmed down, came out of medicinal treatment, and can now, with check-ups every three months, live on her own without going berserk.

"Nikky is not the only case out there," Richardson remarked. He was thinking of all the veterans who have returned home from various wars in the Middle East with serious problems. "Again and again, we hear about them, people who kill their families or themselves. More than a hundred thousand have committed suicide, and even though some of this has to do with psychological traumas, I think many of them have had some brain damage."

Traumatic brain injury (TBI) has been called the "signature injury" for the wars in Iraq and Afghanistan. Repeated head traumas from explosions and other blows harm the brain tissue and yield unpredictable symptoms. It might be aggression, depression, anxiety, a shattered memory, or personality changes. Since 2000, almost 300,000 American military personnel have suffered traumatic brain injury, and more than 2 million returned veterans struggle with neurological or psychiatric ailments. This huge number of sick people has become a major, recurring debate, and their situation has forced the US military to move into brain research.

"The Military Wants to Fix Damaged Brains," read a headline from a 2013 edition of *Science*. That was the year that the Defense Advanced Research Projects Agency, DARPA, which is the research arm of the American defense forces, offered $70 million to develop the next generation of deep brain stimulation and, in the process, to revolutionize our knowledge about the biology of psychiatric illnesses. The initiative got tremendous media attention, and the most enthusiastic commentators compared it to Kennedy's moon-landing program, calling it a *moon shot for the mind*. I heard of the project for the first time from Helen Mayberg. She had applied for a grant but did not get it. She called the ambitions of the military boys "totally unrealistic." At one point in our conversation, Mayberg even used the word "outrageous," and she gave me a copy of the project description to take home, so I could form my own impression.

The goals *were* ambitious. It sounded like science fiction.

DARPA wanted to construct small electrical systems that could be embedded in the cerebellum. A sort of electronic superego. The agency wanted units that not only stimulated selected brain cells within fixed parameters but read the state of the brain on a running basis and corrected it in order to *prevent* certain feelings and certain types of behavior from ever arising. One could talk about a mental *preemptive strike*.

The principle itself resembled something I heard scattered remarks about at the surgeons' conference in Maastricht. It was called closed loop, and the system existed in its initial primitive version for the treatment of epilepsy. One part of it consisted of one or two electrodes, which captured irregular brain activity like a sort of earthquake detector. The other part consisted of electrodes that stimulated the epileptic area of the brain within certain parameters in order to moderate the activity and, so to speak, strangle the fit at birth.

DARPA envisaged a similar approach to an array of conditions from which war veterans and military personnel suffer. In the project outline, they mentioned post-traumatic stress and traumatic brain injury but also depression, anxiety, and borderline personality disorder, as well as drug abuse, fibromyalgia, and memory problems. As Helen Mayberg put it, "In reality, a hodgepodge of maladies whose mechanisms we know only a very little about."

Just take post-traumatic stress. People suffering from it are tormented by gruesome flashbacks and debilitated by anxiety attacks. The DARPA boys imagined and called for an intelligent, implantable system that captured signs that something was on the way, reacted in the space of milliseconds, and prevented the signals from making their way into consciousness. Researchers at Massachusetts General Hospital already have a rough idea about how to do this. They know that fear is generated in the amygdala but also that the amygdala's activity can be modulated and repressed by another higher region—the ventromedial prefrontal cortex. The idea is to detect hyperactivity in the amygdala and, through quick stimulation—for example, in the cortex—create a signal that dampens the amygdala before the fear reaction blossoms fully.

"This is something completely different and new. This apparatus does not exist yet. We are trying to change the game in relation to how you tackle these types of problems," declared Justin Sanchez, who was the project's coordinator at DARPA, to the *New York Times* in 2013. At the same time, the agency's deputy director Geoffrey Ling said to *Science* that deep brain stimulation has traditionally received too little money. "So, we thought, what the hell? Let's give it a try and be bold—if it works, it will be fabulous."

The audacious tone was nothing new at DARPA. The agency was a Cold War invention, founded in reaction to the American shock of the Russian launch of *Sputnik*. It was supposed to ensure that the United States never again limped behind in cutting-edge technology. DARPA is perhaps most well known for its role in the invention of the Internet, which began under the name Arpanet as a development project in the late 1960s. It was also with DARPA that the first Internet pioneers invented the basic IP protocols that today keep everything running. Since then, DARPA has used its powers to breathe life into artificial intelligence, but they have also begun to pump more and more funding into biological and medical technology. This arm of the military is also known for thinking bigger and wilder than any other money bin. As one researcher put it at a public seminar in Silicon Valley in 2015, DARPA is like a "friendly, but somewhat crazy, rich uncle."

I wanted very much to speak to Uncle DARPA, but for the longest time he would not respond. No matter which number I called of the many listed on the agency's homepage, I ended up with the same insipid answering machine. Finally, after quite a bit of maneuvering, I got through to Geoffrey Ling—Colonel Ling—who had just left the agency. He called me back from the road somewhere on the East Coast. One of the things I was curious about with respect to deep brain stimulation was this: They want scientists and researchers to create new systems, use them to map different types of abnormal brain activity, and test them on people—all over a period of five years. Five years! That is a blink of an eye in normal research contexts.

"It is mega-ambitious," conceded the fast-talking Ling, but he assured me this is a very DARPA-esque approach to things.

"This is not the NIH. It was never a matter of 'this is a neat area to study.' We wanted a system with a real capability, and you have got to show it makes people better. Otherwise it's just a science-fair project."

So studies on mice and monkeys were not going to satisfy the military, and the extremely short and rigid timeline was set to ensure that only the "right guys" would apply. As Ling put it: "You want the guys and gals who have got the guts. Who are willing to put up or shut up."

Some of those guys may be found at Harvard University and the affiliated Massachusetts General Hospital. Here, a trio consisting of neurosurgeon Emad Eskandar and psychiatrists Darin Dougherty and Alik Widge joined up with some engineers at the nearby Draper Laboratory to accept $30 million and the five-year challenge.

When I arrived in Boston, Eskandar was out traveling. So I met with Dougherty and Widge in the historic building of the city's old navy yard, where Harvard's neuroscientists were housed together with the highest concentration of brain scanners anywhere in the world. When I walked through the enormous atrium, I could hardly keep from smiling at the irony of the situation. This was where the military stored bombs for

Atlantic convoys during the Second World War. Now that the elite troops of the academy had moved in, this same military was ponying up the research funds. On the other hand, the Harvard group was well on its way toward providing a device that sounded like DARPA's wet dream. In my own mind, I have dubbed it *the electronic superego.*

"That's not far off," said Dougherty, whose cavelike office we were using for the meeting. On the face of it, he and Widge made for a strange duo. The middle-aged Dougherty was fair, round-headed and almost childlike in his expression, while the younger Widge was dark, slender, and intense with a Mephisto-like goatee. They completed each other like an old married couple, and as soon as his colleague paused, Widge nodded and said, "The project *is* wildly ambitious."

More specifically, we were talking about a little computer that could be integrated into a human brain and connected directly to nerve tissue in order to keep the organ on the path of the straight and narrow, so to speak. Through delicate electrodes, the apparatus could measure the brain activity in selected areas on a regular basis and, with electronic stimulation, correct it as soon as there was a sign that something unwanted was on the way. It might be brain activity that indicated a depressive direction or, if allowed to develop, would lead to anxiety, compulsive behavior, or exaggerated impulsiveness.

The apparatus itself existed only as a prototype, developed by the engineers at Draper, and it resembled, most of all, a metallic spider. The computer unit consisted of a central "body" from which five arms extended, and the tip of each arm deployed a very thin electrode. They were to be placed in various spots in the patient's brain tissue, depending on what exactly was to be accomplished. At this point, the device is slightly larger than an iPhone, but the next iteration should be small enough to place inside the cranium. The day of the pulse generator with a battery, which in current systems are located beneath the skin of the chest, will be over.

"Generally speaking, we want to get beyond old-fashioned brain stimulation," said Widge. *Old-fashioned*? "Yeah," continued Dougherty, referring to the two high-profile clinical tests of deep brain stimulation for depression that failed, the ones that Thomas Schläpfer believed had been launched too quickly.

"We participated in the one funded by the company Medtronic right here," Dougherty said. "And we think the reason both trials didn't show any effect, was because the very procedure is based on an outdated mode of thinking about psychiatric illness. A mode of thinking that we are changing radically."

"Radical" is definitely the word. Dougherty and Widge advocate scrapping the very diagnoses on which the whole of psychiatry is based. Highly respected and well-known categories such as major depression, OCD, social anxiety, or personality disorders are inadequate because they do not reflect the underlying physiological reality. Just take the concept of "depression." It sounds clear enough but is actually a rather flimsy thing whose contours become more fluid as you try to define them, because the experiences and symptoms of patients vary considerably. Some sleep too much, others pretty much cannot sleep.

A number overeat and gain weight, while others lose their appetite and waste away. Some have increased levels of stress hormones, while others do not. Among themselves, psychiatrists murmur still more loudly that they are in fact dealing with several diseases that also have different organic causes and biological mechanisms.

At the same time, there was a striking degree of comorbidity within psychiatry. Which meant that the individual patient quite often qualified for a number of diagnoses. One could have OCD *and* depression or one have a personality disorder *and* be riddled with anxiety. Not to mention the many with anxiety *and* depression. Dougherty leaned forward in his office chair to emphasize the implications:

"The symptoms for the old disease categories overlap, and the key to understanding what is going on is to look at the behavioral domains that are touched. In other words, we have to analyze what basic brain processes are disturbed in the individual."

You might just as well grab the bull by the horns and call it a paradigm shift, and it is trickling down from the top of research-based psychiatry. In 2008, the National Institute of Mental Health launched a new strategy. The agency called for projects that looked for new ways to classify patients' conditions on the basis of "dimensions of observable behavior and neurobiological measures." The initiative was dubbed RDoC (for research domain criteria), and it was in clear opposition to the eternally debated *Diagnostic and Statistical Manual of Mental Disorders*. This "bible for psychiatry," which was published in its fifth updated version in 2013, has been criticized for constructing ever more diagnoses on the basis of clinical observations and growing lists of symptoms instead of knowledge about mechanisms.

Dougherty and Widge were inspired by the RDoC and called their own approach a "transdiagnostic framework" because it cuts across existing diagnoses. They have selected six behavioral domains for their analyses. Domains that, in different ways, are touched or disturbed in a number of psychiatric ailments: reward motivation, emotion regulation, decision-making/impulsivity, cognitive flexibility, fear extinction, and learning/memory. In every domain, a person finds himself or herself somewhere on an axis between a low and a high score, and in the middle each axis has a normal range where most people are located.

"The crucial thing is that we know the underlying brain networks of these domains," Alik Widge added. "Which means that, in theory, we can manipulate them very directly with our electrodes."

What they were telling me was that there is no one brain area where depression resides while anxiety refers to another area, and addiction to a third. Any given ailment or condition in any given patient corresponds to a personal constellation of different types of behavior. This can be measured as a combination of the score on the various inner axes. You can imagine how two people today might get the same vague diagnosis of "depression," while, with the new approach, they would get a more detailed characterization of their condition. One person might have a hard time getting motivated and at the same time suffer from a great deal of fearfulness, while another might be characterized by low cognitive flexibility, which is expressed in recurring negative thoughts. This would give psychiatrists something for which they have hungered ever since the discipline was founded—namely, actual tests, something that can be measured like blood pressure or cholesterol numbers.

"Biomarkers," said Dougherty, and mentioned as an example cognitive flexibility. This is the ability to control and regulate one's mental activities and it is often compromised in various ways in mental illness. Patients suffering from depression can get stuck in a train of negative thought and, correspondingly, people with post-traumatic stress find that their thoughts keep returning to their trauma. Both problems can be traced back to poor connectivity between regions in the frontal cerebral cortex and the deeper-lying amygdala. So, in theory, brain stimulation that is directed toward these networks can help certain patients suffering from depression or PTSD. And this can be done with one of the five electrodes in the new device, Alik Widge stressed. The rest can be put to work as needed on other networks with which the individual is having problems.

"It is personalized medicine by way of electronics," he said. "We can define pathology as when someone's cognitive flexibility or emotional regulation is measured as two standard deviations from the average. You have to imagine a future in which you test and quantify these behavioral domains and in which treatments aim at affecting constellations of behavior and normalizing them."

I try. But have to ask where the norm is coming from. And

the answer is that, provisionally, it is an average of a test conducted on thirty-six carefully selected volunteers. People who were themselves the very picture of normality. Individuals who in all kinds of standardized psychological interviews did not indicate a single symptom of mental illness. Again and again, this group was asked to take various domain tests, and now their results constitute the golden mean by which others are measured.

"How about taking the test yourself?" Dougherty asked, looking so happy that I couldn't refuse.

He took out a computer and suggested that we begin by examining my cognitive flexibility. The test was called ARC, and it was like playing an incredibly simple computer game. In the middle of the screen was a series of three numbers of which two were identical. With the correct number key, I was supposed to register which number was the 'odd one out.' The confusion arises in that it might be the number 3, but it was situated as the first or the second in the row—and the worse your cognitive flexibility was, the more often you pressed the wrong key. I went through the entire test and ended up with a score of 100 percent. Darin Dougherty held up a meaty hand for a high five.

The next test assessed one's ability to engage in emotional regulation. It measured a person's ability to ignore irrelevant emotionally charged information when working purposefully on a task. I was presented with faces that had either a fearful or a happy expression. Across each face was the word *happy* or *fearful* printed in red. Sometimes, it was a correct match; other times, the word was wrong. I was supposed to answer based on the expression on the face and ignore the written word. This proved to be more difficult than it sounds. Congruent and incongruent pairs changed randomly and it was confusing. Again and again, I messed it up—at one point slamming hard on the keyboard. Alik Widge tilted his head and looked out the window.

"Yeah, it can be a bit frustrating," said Dougherty. He laughed nervously and closed the computer discreetly. "But our test subjects gradually get used to it, and it doesn't seem to bother them."

Test subjects—I had completely forgotten about them. My immediate thought was: Who in the world could they get to test all these computer tasks with electrodes buried deep in their brains? But apparently recruiting is no problem. There happens to be a group of people who already need brain surgery and are about to have electrodes implanted anyway: epilepsy patients.

"We already have them wandering around the ward. They're bored, and many would rather participate in an experiment. Actually, roughly three out of four say yes when we ask," Widge explained. And in fact, neurosurgeons have been experimenting for a long time with epileptics—people who were quite severely affected and did not respond to medicine, people you tried to help by removing the tiny areas of the brain that gave rise to the fits. They were typically hospitalized and had temporary electrodes inserted in different places in the brain in order to demarcate the affected areas. Then it was just a matter of waiting around for the fits to reveal themselves to the doctors. But if the deep electrodes were already implanted, why not use them to gain some new knowledge about the brain? It would be unethical to let the opportunity go by. At least, so argued a group of neurosurgeons in a 2009 paper for *Nature*.

"Let's see a patient video," Dougherty suggested with a clap of his hands. He quickly found a bit of film with a youngish woman sitting in a hospital bed surrounded by various devices. She had twelve electrodes placed in her brain, and each of them had ten different contacts from which to take measurements. This setup reminded me about how witnesses at the Congressional Hearing in 1973 accused Robert Heath of turning his patients into "human pincushions," and I asked cautiously whether any damage might come from it. Dougherty shook his head.

"The good thing about brain tissue is that it's kind of gelatinous. It's like putting a thermometer in Jell-O. It pushes everything aside—okay, it shears a little bit—but when you pull it out it all just kind of retracts and goes back to normal."

Alik Widge continued his explanation that the patient was hooked up to the first version of the provisional system, which was external and took up as much space as an old-fashioned stationary computer. First, the person underwent one or more tests, while the equipment measured the networks in which the researchers were interested. Gradually, they mapped the individual's initial condition and then calculated algorithms for how to do the stimulation to affect the tested behavior.

In the video, the test was of cognitive flexibility. Alik Widge was sitting outside the picture, chatting amiably with the patient. Just like an updated version of Robert Heath's old black-and-white footage.

"Do you feel anything in particular?" we heard him say.

"So . . . um," she answered. "When I'm usually doing the

task, I'm, like, counting the numbers and then figuring out which one I'm searching for. Right now, I'm feeling a bit like it just comes automatically."

"An increased sense of flow in some way?"

"Yeah, a sense of being in the zone and automatically knowing exactly what to do."

Widge stopped the film and remarked that this patient had a very anxious style, "close to an OCD diagnosis," and that she always scored 100 percent correct on the numbers test.

"But you can see that the stimulation increases her cognitive flexibility in that she completes it far more quickly than normal. This is actual improvement of decision-making ability."

At that point, six test subjects had tried out the electronic superego, and the apparatus and its algorithms did what they were supposed to do, pushing behavior in a given direction. With their equipment, Dougherty and Widge could make a person less impulsive or more cognitively flexible. Just as they could improve a person's emotional regulation or actually give them more affect if they were emotionally numbed out.

Now it was just a matter of getting the engineers at Draper to shrink the prototype a bit more, so it could be inserted surgically and then wait for the first clinical trials for which the FDA had already given the green light.

"When will it be?" said Widge. "Sometime in 2019, if everything goes according to plan. And we hope it will."

I was fascinated. Not only by the advanced projects in all their futuristic glory but especially by the historical parallel, which screamed out to heaven. In the 1950s, it was the CIA that used brain research; today, it is DARPA. And then, like now, top scientists were involved—and happily. To judge from reactions in the media, only a minority of today's researchers have anything against military funding. Steven Hyman, who heads up psychiatric research at Harvard's Broad Institute in Boston, told the magazine *Spectrum*, "The kind of hardware that DARPA is interested in developing would be an extraordinary advance for the whole field." And when at one point in my conversation with Darin Dougherty and Alik Widge I asked whether they had any reservations about receiving funds from the military, I received a convinced "not at all." To the contrary, they were quite happy about it. The DARPA grants were part of President Obama's Brain Initiative and, after all, the Harvard group were developing a medical technology, not at military one, as Widge put it. In addition, DARPA established an ethics panel to monitor the research and avoid difficulties with the public.

It could also be added that whereas the CIA in its day wanted to get methods for coaxing information out of enemy soldiers, DARPA only wants to help its own returned veterans—and who could be against that?

But the objections are obvious. Once the technology exists and is ready to be implemented, it could be used in many ways. If a chip in the brain is capable of moderating a post-traumatic stress reaction, it could presumably also prevent the horrific experiences on the battlefield from even creating psychological trauma. And could you imagine using these intelligent implants to make soldiers tougher and, thereby, more battle-ready before you send them out? This might even sound like concern for the individual soldier. By the same principle, you could also manipulate a brain to be more aggressive or callous.

In fact, the first experiments to affect ordinary yet complex personal traits have already been carried out. They have shown, for example, that research subjects can be made more persistent with electronic influence and that their relationship to social norms can be changed.

One 2013 study published in the distinguished journal *Neuron* described two epilepsy patients who were part of an experiment at Stanford University and subjected to a mild current through a single electrode placed in the anterior midcingulate cortex. To the experimenters' surprise, the manipulation gave rise to the exact same reaction from two otherwise very different people—namely, the feeling of a huge and persistent drive. A feeling that something had to happen, that something had to be dealt with. In other words, a little jolt of electricity could arouse huge motivation without a specific object.

That same year, economist Ernst Fehr of Zürich University experimented with external electrical stimulation. What is called transcranial direct current stimulation sends a weak current through the cranium and is able to influence activity in areas of the brain that lie closest to the skull. Fehr had sixty-three research subjects available. They played a money game in which they each were given a sum and had to take a position on how much they wanted to give an anonymous partner. In the first round, there were no sanctions from the partner, but in the second series of experiments, the person in question could protest and punish the subject. There were two opposing forces at play. A cultural norm for sharing fairly—that is, equally—and a selfish interest in getting as much as possible for oneself. Fehr and his people found that the tug of war could be influenced by the right lateral prefrontal cortex. When the stimulation increased the brain activity, the subjects followed the fairness norm to a higher degree, while they were more inclined to act selfishly when the activity was diminished.

Perhaps the most thought-provoking thing was that the research subjects did not themselves *feel* any difference. When they were asked about it, they said their idea of fairness had not changed, while the selfishness of their behavior had changed. Apparently, you can fiddle with subtle moral parameters in a person without the person who is manipulated being any the wiser.

Of course, we're only talking about initial results and isolated experiments, but they call attention to something crucial. With electrical stimulation, we have put our hands directly on the reins of human nature. In principle, there are no limits to the tendencies and qualities that can be modulated. It seems we can make people into tools to do anything, and that emotions and ethics are malleable things.

Yet, we are not talking today about *mind control* in the concerned way they did in the 1970s. You have to go to niche publications like *Catholic Online* to read about DARPA's projects with the perspective that "such tools could theoretically be used by dictators and despots to control their population." The words sound almost antiquated. You could say the same about the statute they passed in the state of Oregon in 1973, which prohibited any form of psychosurgery "whose primary purpose is to change a human being's thoughts, feelings, or behavior." Today, the statute is being reconsidered now that deep brain stimulation is rising meteorically as a promising treatment method—a method, take note, that has the *express* purpose of changing feelings, thoughts, and behavior.

The thing is that, when we now hear a phrase like "mind control," it doesn't sound like something spooky, like alien powers pulling the strings. It sounds more like being in control of your own mental equipment. It is about shaping yourself in accordance with your own needs and ambitions—actually, just an extension of everything else we do to optimize our ability to meet life's demands. Yoga, mindfulness, and various brain training programs you engage in with a computer sound more harmless, of course—some will say more "natural"—but they are all fundamentally directed at changing your brain processes.

My fleeting insight into DARPA's accelerated projects made me dizzy. It was the gap between Robert Heath's homemade electrodes and today's electrode blankets with thousands of microscopic feelers that hit me, along with the gap between the people's attitudes then and now. I felt just like the time, many years ago, when I visited the Air and Space Museum in Washington, DC. I stood in the museum's great hall with the Wright brothers' rickety flying machine on one side and a shiny space capsule from the Apollo mission on the other. I kept turning my head from one construction to the other, thinking how inconceivable it was that there were only sixty years between them.

Soon after my almost futuristic encounter in Boston's navy yard, I was pulled six decades back in time by a package that turned up in my mailbox. It was from Charles O'Brien, Heath's former student, who was now a drug-addiction expert in Philadelphia. In the package was a stack of yellowed papers held together by a plastic spiral with the title and author on the front page in bold, black type: *Psychosurgery—1954 by Arnold J. Mandell.*

It was the manuscript of Arnold Mandell's "novel," with a cover letter to the colleagues to whom he had originally sent it.

> I wrote this in the middle of the 1960s from notes taken while I was a medical student at Tulane's psychiatric ward between 1954 and 1959. The experience was so overwhelming that I see it as an attempt to process what it really was I had been a part of.

Of Heath, Mandell wrote: "I was, like most others, dazzled, inspired, and confused by his breadth and charisma."

He apologized that the text was "not especially well-written" but stressed that "everything in this book actually happened."

I put the letter aside and ran a finger down the table of contents. I was looking for clues to what Charles O'Brien might have thought of when he mentioned that Arnold Mandell could have the key to understanding what happened to Robert Heath.

As I dug into the pages, what I found was not only completely unexpected but had a touch of the bizarre. Mandell took me back to the late 1950s and a time when Heath was up and coming and in the headlines. But it turned out that it was not his forays into deep brain stimulation that precipitated his downfall. Rather, at the heart of the matter seemed to be a different line of research, one that I had heard very little about and which I had all along viewed as just an insignificant aside.

CHAPTER 9

A Grand Mistake

Where the hell was Irene?

The afternoon was turning into evening, and Robert Heath had turned on the desk lamp while attending to a pile of papers in front of him. Administration was the part of the job he liked the least. Normally he hurried through it, just signing whatever Irene put in front of him. But today, something caught his attention. Something was very wrong. He turned three invoices this way and that. They had been issued in his name and were recently dated—May 1957—but he did not recognize them. A car had been rented, which he never drove; men's clothing had been purchased from an expensive tailor he did not care for; and several dinners had been eaten at the fashionable Galatoire's restaurant, a spot he had not frequented in months.

Heath could make no sense of it. Irene did not understand it either. These were hefty bills that, sure enough, bore Heath's signature, but it had not come from him. The two debated back and forth whether the slanted *e*'s, which were rounder than his own, or rather the large *R* revealed the forgery. But who would do this sort of thing?

Heath sat in the circle of light around his desk while Irene paced the floor, going through the possibilities. They ended by calling the tailor's shop and talking to an employee. Off the top of his head, he could not remember the invoice about which they were inquiring but offered to rummage through the company records and get back to them as soon as possible.

When the telephone rang again, it was with a strange piece of information. Yes, one of the trainees remembered the order for that particular set of clothing. It was made by a man who claimed he was buying it on behalf of Robert Heath. A very short man it was, with a distinct New York accent.

Heath put down the receiver. He now knew the matter had to do with Matthew Cohen. The way Irene furrowed her brow was familiar to him. Cohen was a biochemist in Heath's group and something of a personality. He was dark and muscular but so short that he put cork insoles in his shoes to gain an inch or two. They laughed about that at the lab but always on the sly. Because there was something intimidating about Matt Cohen. He was not only strikingly gifted and the type who quickly dominated any conversation in which he participated but was also fierce and eccentric. He was especially fascinated by guns, and Charlie Fontana had seen him carrying different ones in the lab. Recently, Cohen had come to work with a gunshot wound in the leg, which he displayed with a strange smile but explained no more about it. And several times Charlie had seen him take money from a thick roll of dollar bills in his pocket.

Matt Cohen had been instrumental in Heath's most promising discovery to date. Something he hoped would be his primary contribution to research—perhaps even the solution to the mystery of schizophrenia. Heath had named it taraxein.

The expression came from the Greek word for disturbed mind—*taraxis*. Heath coined "taraxein" to name a proposed substance swimming in the blood of his schizophrenic patients. He imagined that the substance was a protein, and one that somehow disturbed the transfer of information by nerve cells in some way and was behind the characteristic explosive activity he was measuring in the septum of schizophrenics.

He had recently done some experiments that made colleagues around the country sit up and take notice. With protein taken from the plasma of schizophrenics, he had on several occasions produced temporary psychotic symptoms in normal research subjects. This was exceptional, dramatic, and indicated that the cause of the disease could be found in the blood. Now, staring at the forged bills on his desk, Heath got a chilly feeling that something was very wrong.

In a few months, both he and Irene knew they would be put to the test. Somewhere on the desk in front of him lay the invitation to a Macy conference, which was an informal conference with huge prestige but also great risk. Every year, the wealthy Macy Foundation held an exclusive meeting in which they scrutinized a handful of the year's most sensational medical findings. The format was designed to confront and test the tenability of scientific innovations, and they summoned the brightest professionals around to conduct the examination. Heath's mind was frantically running through everything he knew about Cohen. One thing that hit him was how Donald Richardson, the young surgeon, had mentioned to him how Cohen always shut himself inside the cold-storage room for hours to make his preparations. And that he forbade anyone from coming in. Had he, the man in charge, shut his eyes to something he really knew he should have investigated?

What should he do? He knew that there was a whole slew of critics ready to flay him and his team, and he could not possibly ask for a postponement—it would sound suspicious. And he couldn't say that the reason was that he suspected his biochemist of—of what exactly?

He had to find out what was going on with Cohen and whether he had somehow compromised the precious protein preparations. He felt sick at the thought that this groundbreaking discovery might prove to be nothing. He needed Cohen to explain himself.

According to Arnold Mandell's convincing account, the meeting was contentious. From the front office, Irene listened to the conversation through the synthetic filter of the intercom. At first, Heath held his rage back and tried to retain the advantage with a restrained but insistent tone. Cohen, who had come straight from the laboratory, sounded cool and relaxed, as if it was something he had been expecting and was ready for. What did Bob mean by "who is he really"? Does it even matter—shouldn't his work speak for itself? Science is like art in that the creator's private life does not matter for the work. And might he not have produced the group's most significant discovery so far?

But people cannot reproduce the results, objected Heath in a low voice. Several groups had tried, but they kept getting negative results. And they were not incompetent people. Cohen smiled and said that he could understand his colleagues' problems, because he had omitted a crucial step in the complicated purification process in his description in their publication in the *American Journal of Psychiatry*.

"What?" screamed Heath in a strange falsetto. "*Omitted?*"

The whole purpose of describing your methods down to the most insufferably trivial details was so that anyone could replicate them. Replication was the very foundation of science!

Very possibly, responded Cohen calmly, but he thought of his little secret as life insurance. As long as there was one technical quirk that only he knew, Heath could hardly get rid of him even if he found out who he really was.

As Arnold Mandell's account further has it, there was a muffled silence on the intercom, which was broken by the sound of a roar. Heath's frustration got the better of him, and he allowed himself to give vent to his temper. "You are a *fucking* psychopath," he shouted again and again at the silent Cohen. "Goddamn liar!" Did this scumbag have any idea what he had done? It was not just his, Robert Heath's, whole career on the line but the entire university department's, with dozens of employees for whom he had responsibility. Not only that but biological psychiatry itself and the credibility of everything they had been doing!

Yes, well, Cohen could disappear if that was what Bob wanted. He could go back to where he came from, he said in a derisive tone. Then the entire story poured out of him. Outside in reception, it sounded to Irene as if Cohen almost got excited as he was telling it.

It was like Cohen's large family was divided right down the middle by an invisible line. There was the respectable side, where everyone was a doctor or a lawyer, and there was the other side that cultivated Mafia connections and lucrative illegalities. He, Matt Cohen, had a cousin about the same age with the exact same name on the other side. The cousin was a trained biochemist from Yale. "With a PhD and everything . . ."

Cohen stopped for a moment and, during the pause that followed, both Irene and Heath drew the same conclusion on each side of the door. The man they had let run the laboratory with a number of assistants under him was not even a scientist but a convincing swindler.

Cohen continued. He talked about the shady family business he came from. Until two years before, Cohen, together with his father and brother, had been part of a successful, international gambling racket, an operation the three Americans ran with some colleagues in Costa Rica, skimming off the cream. The money was flowing, and the not quite thirty-year-old Matt was doing just fine with a penthouse apartment on Miami's harbor, an opulent house in one of the city's most expensive suburbs, and more luxury cars than he could keep track of.

"But this thing runs fast," he said, tapping his temple with a finger, "and I started to get bored."

The young petty gangster had always been fascinated by science and, in his boredom, he had read about Heath and his electrode studies, which had captured his attention because his sister was in an insane asylum in New York. She was diagnosed with schizophrenia, and he had always been interested in the psychotic. So, why not just pay a visit to his Yale-educated brown-noser of a cousin and sneak a copy of his business card? It worked like a charm.

Heath was crestfallen. He could understand how Cohen might get hold of forged diplomas. But he could not understand how he could pull the wool over everyone's eyes so effectively—he had dazzled with his familiarity with the professional literature and the way he handled the technology.

That was easy to explain, said Cohen. In the months before he began at Tulane, he got a job at a biochemical lab in California—at first, he washed the equipment, but he quickly worked his way up to become a technician and learned what was needed. Risky, he admitted, but not at all impossible.

"It's not like there are qualified biochemists flocking into psychiatry, is it? I figured it would be easy . . ."

Taraxein was Robert Heath's baby, the offspring of his own theories that had taken shape with his experiments. The idea of a biochemical abnormality that manifested itself in schizophrenic brains pulled together all the threads he and others had derived from their observations. Taraxein seemed to tie things up in a logical and beautiful knot.

First, there was the electrode stimulation. When, early in the

1950s, the Tulane group discovered the chaotic activity in the septum of schizophrenics, Heath believed that the tissue in the area must have been compromised. It was known that there were no visible anatomical deviations, so it must have been a result of something abnormal in the transfer of information to and from the cells in that small area of the brain. As he said, "There must be something wrong with the synapses." And even though the detailed mechanisms were not well understood, it was clear that the biochemistry had to be investigated.

Things were already happening. A handful of biochemically oriented psychiatrists were searching the bodily fluids of the insane—blood and urine—to find a trace of something that distinguished them from the normal. They had an intuition that there *must* be something different and that it was just a matter of using the right method to reveal the culprit. In 1952, Abram Hoffer and Humphry Osmond claimed that schizophrenia was the result of an odd metabolism of a particular group of transmitter substances—the catecholamines. This inspired Heath to settle on the protein ceruloplasmin, an enzyme thought to metabolize certain catecholamines. What if there was more or less ceruloplasmin in schizophrenics?

In 1956, Heath and his colleague Byron Leach found that the metabolism of catecholamines took place quicker in schizophrenics than in normal people. But the trail led to a dead end. As soon as they investigated people with a number of other diseases—including the common cold—it was clear that schizophrenics were not special in that respect.

Therefore, ceruloplasmin could not be the villain. But what if there were something *else* in schizophrenics, Heath theorized—an enzyme that accompanied ceruloplasmin when they did the purification but was actually different? And what if that was what produced the symptoms?

He pursued the case by taking blood from his schizophrenic patients—liters of it over time. As the person responsible for biochemistry, Byron Leach went about purifying and concentrating protein fractions, which they could then inject into monkeys. The animals, which already had permanent electrodes imbedded in the septum, were to be the researchers' measuring equipment. They were on the lookout for changes in their behavior that might resemble schizophrenia and for the explosive activity in the septum, which was the signature of patients with acute schizophrenia.

But at first very little happened.

Charlie had given the injections, and Heath sat in front of the cages with all his senses primed and a hope that almost burst from his breast. Any little change in the monkeys' conduct was pointed out, discussed, and noted.

"Doesn't he look a little nervous? Isn't he more aggressive than normal?"

But it was never very convincing. Leach was well respected, and he had run every dodge to take care of the biochemical end, but after some months there had been no real breakthrough, only hints of an effect. Then, Matt Cohen entered the picture. One day, he showed up to meet Heath, put a stack of papers in front of him, and flattered him with statements about how he admired the work at Tulane and wanted to use his life to contribute. He also said with great self-assurance that he knew all about proteins.

It worked, and after a couple of calls to the academics he listed as references, the man was hired. Good biochemists were hard to find—especially if you were in psychiatry. So you had to strike while the iron was hot. Quickly and without anyone being able to explain why, Cohen came to dominate the biochemistry in the group and had a whole slew of people working for him. However, there were certain things he did not delegate to others.

But after a few months of work, he gave Heath what he dreamed of. Finally, one day, the monkeys reacted with something that looked almost like catatonic symptoms. Anyone could see that something was afoot—just a few minutes after his injection, the first monkey looked petrified in his little chair and did not react to much of anything. The technician and jack-of-all-trades in the group, Herb Daigle, stepped in with his usual cigarette in his mouth and stroked the little animal cautiously on the cheek, then on the head. It was completely gone. It did not even blink. Then he took a hand, then a foot, and stretched the limb out and let it fall. The monkey was completely relaxed and stared into the air with big, round eyes. Incredible! Shortly thereafter, a cheer went up in the laboratory when Charlie Fontana caught some septal hyperactivity on his EEG strip.

Eureka! They had done it, and the feeling was intoxicating.

Then, it was time to contact Maurice Sigler, the director of the state prison in Angola, to hear whether it might be possible to test the new serum on volunteer inmates. Sigler believed it could be done. And with approval from the district attorney in New Orleans, they were presented with a consent form.

It was never hard to get volunteers, because anything that broke the tedium at the notorious, barbaric prison was welcome. Heath first chose two suitable petty criminals—young white men who had been examined and seemed to be psychologically normal. It was important to avoid any deviant characters. A guard drove the two prisoners to Tulane and, by turns, they were seated in the chair in the film room on the second floor, where Heath spoke to them in front of the camera. They chitchatted a bit about prison life—Ed, as the first prisoner was called, worked inside the walls as a barber. Before the injection, whose contents they did not know, their sleeves were rolled up. The two were each injected with the preparation that had already been shown to affect the monkeys. The milliliter and a half went quickly into their veins, and not more than a minute went by before things began to happen.

"This is *exactly* like our first monkey," remarked Charlie, who was one of the flock of observers behind a one-way mirror. On the other side, Ed dropped his burning cigarette without reacting. His eyes were open but did not seem to be working, and his head was lolling around his slender shoulders as if he was having a hard time holding it up. Heath slowly picked up one of Ed's arms with rolled-up sleeves. There was no reaction, and it just went slowly down again when he let go. The young man did not answer when Heath spoke to him except for a single time, when he emitted a small mumbling sound.

Dean's reaction was livelier. He reacted to his injection almost with paranoia. He squinted his eyes and looked at Heath as if he were an enemy one had to be wary of. "They're talking about me out there," he said angrily, without explaining who "they" were or what "out there" meant. "Everybody is talking and talking. Get me out of here!"

Two experiments: One became catatonic, while the other became paranoid—it was almost better than they could have hoped for! The two clearly affected prisoners were Heath's trump card. Now they could go out with everything they had. In May 1956, the bigwigs from the Tulane group went to Chicago, where the American Psychiatric Association had their annual conference. On the last evening, during a celebratory dinner, Heath laid out their data in a session dedicated to cutting-edge research. As expected, the two film clips made their colleagues' jaws drop.

"Dramatic," a reporter from the *New York Times* called the experiment, and the leading US newspaper used column space on the story two days in a row. Because if this mystical substance that had not yet acquired a name really existed, it meant that there was a biological cause for schizophrenia—which is to say, something tangible that you could work to find a cure for.

Robert Heath kept his pot boiling. Just a few months later, he was at a conference in Montreal, where he followed up with more experiments. In 1957, the group published an article in the *American Journal of Psychiatry* in which they described the results of twenty volunteers. Here they also unveiled the name of their postulated substance—taraxein. This substance, they claimed, produced clear schizophrenic symptoms in all the research subjects, whereas corresponding protein preparations taken from normal subjects produced no reaction at all.

But not everything was rosy—far from it. The discovery of taraxein aroused not only interest but acted like a red cape for some of Heath's colleagues and competitors in the field of biological psychiatry. They did not trust him. At the presentation in Chicago, the hostility began with a remark from Douglas Bond, the respected head psychiatrist at Case Western Reserve University in Cleveland. This bald gentleman put down his napkin, cleared his throat, and calmly said, "A bunch of psychiatrists do not need to be reminded that the easiest people to fool are ourselves."

A year and a half later, in 1958, the discussion continued in an auditorium in Princeton, New Jersey, where the prestigious Macy conference took place. Outside, it was a warm Indian summer, but inside, the thirty-two researchers needed no air-conditioning. The atmosphere was decidedly chilly. They were well-known specialists, who had come to pick apart the taraxein story and put the blame on Robert Heath and the two biochemists he had brought with him. No one there knew what had been going on in New Orleans, and even back at Tulane everything had been kept under wraps since the confrontation with Cohen. During the intervening months, Heath had arranged for a lab technician to follow Cohen's goings and comings in the laboratory, even though Cohen had cut back severely on his contact with colleagues.

Now, in the auditorium, Heath gave Cohen a brief glance before he rose to introduce the proceedings. He sketched out his ideas and the principles in his experiments, then showed a couple of films of selected research subjects. As the first film ran, the room watched in total silence a thin man with slicked-back dark hair and crude, light institutional clothes.

"*I hear something . . . voices*," he said from his chair in the middle of the picture. His eyes had a glassy sheen, and his face was as stiff as a mask. Every so often, he moved his lips without saying anything until a few hesitant words came out.

"*The voices . . . they're spooky. Like something from . . . I don't know . . . like in* Frankenstein," he said to an interviewer outside the picture.

With a pointer in his hand, Robert Heath explained how the young man—being punished for theft—also complained that he could read the interviewer's thoughts along the way. The effect disappeared gradually, and an hour and a half after the injection, there were no symptoms. "The subject was a little concerned but expressed a willingness to receive another injection later that day," explained Heath.

This time, it worked differently. After just fifteen minutes, it was like he was petrified. He stared emptily in front of him, while the cigarette between the fingers of his right hand was allowed to burn down to the skin—to which he paid no heed. Then the interviewer—Robert Heath himself—came in from the side and began without explanation to tug on the man's arms. He pumped them up and down to illustrate how the limbs were slack and malleable. Then he removed the cigarette butt and raised the right arm over the man's head. It remained where it was. "Lower your arm," he said, but nothing happened. The arm remained stiff, at an odd angle, and the man just stared straight ahead.

As soon as the black-and-white film was over and the lights were brought back up, the assembly had at it. The chairman, Hudson Hoagland, from the Worcester Foundation for Experimental Biology, attacked harshly.

"Couldn't it have been hypnotic suggestion?"

Robert Heath did not think so, and referred to the fact that the man burned his own fingers without noticing it. But Hoagland insisted: "How do you know that he burned his fingers?"

It continued from there. There was a widespread and clearly expressed suspicion that the volunteer subjects were performing because they sensed what the researchers wanted to see. But these were blind experiments, argued Heath, in which neither the subject nor the interviewer had any idea what the injection contained. But was that true? Many of those present seriously wondered whether it might have been possible to see if the injections contained saltwater or something stronger. As Hoagland said, the prisoners knew perfectly well what the experiment was about: that they got a drug that could produce unusual effects.

Undeniably, they did, admitted Heath, but no one had the same schizophrenic-like symptoms when they received substances that clearly affect the psyche. For example, the LSD they had all done experiments with. Nor did anything happen when they received an ineffective serum from normal, non-schizophrenic donors. And there you would expect an effect if it were only suggestion or acting.

Ah, but precisely the *absence* of a placebo reaction was in itself highly suspicious, claimed Harold Abramson from the Cold Spring Harbor Laboratory. There was *always* a placebo reaction in medical experiments. Even when you gave normal control subjects salt tablets, some reacted with rashes or other bizarre effects. He saw that sort of thing all the time in his own experiments, and if you did not, the experiment must have been manipulated, he argued.

"Maybe, your normal controls just aren't normal," Matt Cohen suggested cheekily. But he was brushed aside. He was the only one participating in the program who was referred to as Mr. instead of Dr., as someone with a PhD degree would be.

The participants asked again and again to squeeze out the last detail of how each individual phase of the experiment took place. Of course, there were the purification procedures themselves, and what they even got out of the blood of their sick patients. And what about the patients they had taken blood from? Were they people who had met the same requirements for a diagnosis of schizophrenia, or was Heath's definition of the disease too broad?

The discussion swelled back and forth with attack and defense. But it always came back to the greatest weakness in the case: the fact that no one had yet been able to reproduce the findings from Tulane. And a number of laboratories had been trying. In particular, Eli Robbins, who was a rising star at Washington University in St. Louis, had been making persistent efforts. Right after the first announcements, he went to Tulane to learn the purification procedure and then went home and recruited fifteen volunteer research subjects—also prisoners.

But as he demonstrated that day, the prisoners from Jefferson City State Penitentiary in Missouri experienced none of the symptoms that Heath had observed. They pumped the prisoners' veins full of the supposed taraxein as well as saltwater, serum from normal people, and even taraxein produced at Tulane. Of twenty-two experiments, there was no reaction at all in seventeen. Five times, the research subjects had believed they could feel some vague sense of being "affected" by something or other, but only three of them had actually received taraxein.

Robbins brought a film along too—silent because he, unlike Heath, could not afford sound equipment. He showed how a prisoner might stare for a conspicuously long time at his arm after the injection but never broke out into the radical peculiarities they had seen in Heath's volunteers.

Then Byron Leach, Heath's trusty biochemist, took the stage, and a long, technical discussion ensued on the extent to which the preparations had been done correctly. It was notoriously difficult to isolate taraxein in an effective form, and even at Tulane it succeeded just a little over half the time. They went into everything from the water's pH level in different places in the country to the type of centrifuge they used in different laboratories.

"I must say that the biochemical data Dr. Heath has reported are still preliminary and leave me considerably confused and somewhat unsatisfied," Seymour Kety finally said. He was the head of the newly established National Institute of Mental Health in Washington, DC, and Frank Ervin called him "Bob's bête noir." The two had been enemies since Heath began at

Tulane, but nobody knew what was at the bottom of it or could remember any particular occasion for it. Perhaps it was just that they instinctively did not like each other. Funnily enough, they had the same roots—they were both born in 1915 in Pennsylvania, and both were strong supporters of biological psychiatry, but beyond that, they were total opposites. Where Heath was charismatic, handsome, athletic, and highly extroverted, Kety was a small man with a slight limp due to a bad foot and "pretty much chronically depressed," as Ervin put it. Where Heath's research was bold and full of ideas—to put it mildly—Kety stood for a more cautious and always extremely thorough approach. It was a clash between a dreamer and a bookkeeper.

In front of the assembly, Kety called the day "an overwhelming experience" but did so with a tone that revealed that he did not for a moment believe what had been presented to him. Then he guided them back to the original point about expectations and playacting. Can it be completely ruled out that the participating research subjects had been told that they would get a new and interesting drug that might give them schizophrenic symptoms?

After a long day, the moderator, Frank Fremont-Smith from the Macy Foundation, took the floor and thanked Heath, Leach, and Cohen for having shared their data. He admitted that the meeting had been "more difficult than usual," but that this was probably because of the exceptionally great potential in the results Heath had presented.

"Obviously, if you are on the right track, Dr. Heath, and we all hope that you are, then you have made a discovery of great importance to medical science and to the world."

• • •

Taraxein, it turned out, was a mirage that Robert Heath would continue to chase. A glorious vision he reached out for throughout his career. At the same time, it became his scientific doom. It demolished his prestige and credibility in leading professional circles.

After the Macy conference, he was once again left with the same assessment that was thrown in his face at his hopeful schizophrenia symposium in 1952. A group of prominent colleagues derided his ideas and believed that the results he presented were a sort of self-delusion—nothing more than wishful thinking and spellbound research subjects who wanted to give the nice doctor what he wanted.

Back home at Tulane, many wondered why Matt Cohen had disappeared overnight. It was as if he had never existed. Bob and Irene did not mention him, and there was no explanation of what happened.

For the first time, Charlie Fontana experienced his boss in a state that approached something like depression. The man who otherwise never bothered about what the world or random people thought about him suddenly seemed insecure and despondent. Especially after he came back from conferences, he would be down in the mouth for days until he found his way back to his usual form. Fontana took it to heart. He had *seen* taraxein preparations work, particularly in the monkeys that he himself worked with, and he never lost faith in the boss's project.

Heath felt persecuted. In 1959, Seymour Kety published two articles in the prestigious periodical *Science* in which he turned the current biochemical theories on the cause of schizophrenia inside out. And he used a lot of column space on taraxein. In the elegant but dry prose that was his hallmark, the powerful head of the National Institute of Mental Health maintained that neither the supposed substance nor its supposed effects had been substantiated.

By now, the whole field knew about their deep enmity and Kety's distrust of everything Robert Heath did. Only a very few also knew how effectively he prevented Heath from getting public research funds. His applications were futile no matter what they were for. The spigot was closed. And after the Macy fiasco and the article in *Science*, it was not long before the Commonwealth Foundation shut its coffers too.

But Heath would not admit defeat. He simply refused to believe that taraxein was just a mistake in the experimental procedure. He had seen it work again and again, and he had interrogated Matt Cohen thoroughly. He witnessed the man swear that he had added neither LSD nor anything else. Before he asked him to disappear, Heath made him work closely with Byron Leach and go through all the procedures very precisely. With them, the laboratory itself could now produce the efficacious protein fractions. Some of the time, at any rate. And as Heath always took care to point out, there were also the Swedes who had seen activity. Swedish psychiatrist Sten Mårtens, who had been a fellow in the Tulane lab for a while and who coauthored the original taraxein publication had later gone back to Stockholm to prepare protein fractions of his own. He had tried them out on three volunteers—staff members at the prestigious

Beckomberga psychiatric hospital—and in 1959 he reported in the literature that two of them had reacted with symptoms resembling schizophrenia.

Somewhere deep inside him, Heath just *knew* he was onto something, and it was just a matter of sticking with it. He had to keep his nose to the grindstone and pay no attention to what others said or thought about it. Because who were they? Envious people with petty ambitions and narrow conceptions of the world.

So he continued. Alongside the electrode projects, which were growing and developing, and the tests of new medications, Heath carried on in relative obscurity, pursuing the idea of taraxein. People in the department got used to seeing a parade of experts in biochemistry transit through, in perpetual search for the mystical protein. It had to be possible to winnow it out from the many ingredients in the biochemical soup that splashed through our brain or, at least, to get closer to pinning down its identity.

And finally something happened. Initially, it turned out they had gone about the whole thing in the wrong way. But after much frustration they had corrected their thinking and found a more promising path. In 1967, Tulane issued a four-page press release that drew attention to Heath's schizophrenia protein. Everything indicated that taraxein was not an enzyme, as they originally believed, but rather an antibody, and that it bonded with the brain tissue of schizophrenics. In other words, schizophrenia was presumably an immunological problem and perhaps even an autoimmune disease in which the patient's own immune system attacked the body from within.

Three years earlier, from purely theoretical considerations, the British psychiatrist P. R. J. Burch mentioned the possibility, and then came Heath and two colleagues with three publications in the same volume of the *Archives of General Psychiatry*, describing experiments that pointed in that direction. The young neurologist Iris Krupp had experimented with new methods in which antibodies in tissue could be detected fluorescently and thus be seen directly under a microscope. When she tested for antibodies in the brain tissue of deceased schizophrenics and normal control subjects, only the schizophrenics lit up—especially in the forebrain around the septum. She also found a similar antibody in a serum from living schizophrenic patients that was not present in control subjects. Finally, Krupp produced special antibodies against tissue taken from the septa of human beings and injected them into Heath's Rhesus monkeys—which, then, displayed catatonic symptoms and an altered EEG.

It was a logical and almost beautiful explanation. As a result of these experiments, Heath propounded a new taraxein hypothesis. As he saw it, schizophrenics—for genetic reasons—produced antibodies against their own septum and that these antibodies engaged this important area of the brain, thus disturbing its electrical transmissions.

It all seemed to hang nicely together, and Robert Heath was ready to confront the world again. In 1967, he hosted what the Tulane group itself dubbed *The Medicine Show*, in which the doubters could come to visit and see with their own eyes what taraxein did. He invited twenty or so colleagues, who were led by the respected psychiatrist Heinz Lehmann from McGill University. Lehmann was a pioneer with the first major studies of chlorpromazine, and he was widely recognized for his neuropharmacological and clinical expertise.

All the stops were pulled out for them—a big dinner in which the liquor flowed, entertainment in the city, nothing understated. Only the next day did things become serious. They met on the second floor, and everyone except Lehmann took a seat behind the mirror with a view of the little test room. There sat a volunteer from Angola Prison and psychiatrist Joseph DiGiacomo, who was not involved in the research and therefore given the job of injecting the notorious taraxein. He did his job and remained sitting there while Lehmann conducted the interview.

It did not go according to plan. Oddly nervous, their inmate looked around, toward the door now and then; he seemed a tad affected, but it was not the psychotic expression the Tulane people normally saw.

"Yes, well, this isn't working," said Robert Heath at one point, losing his cool in the next room. "He is asking the wrong questions!" he said, rushing in to Lehmann. Then he himself began to ask the volunteer whether he heard anything strange, whether he felt persecuted by anyone, and whether that was why he was always looking at the door? From behind the mirror, the group of visiting colleagues cast glances at one another.

A few days later, after everyone had left, Jim Eaton was sitting in his office when his boss and mentor knocked and swept in. "How do you think it went, Jim?" Heath asked in a muted tone of voice. "Tell me how I seemed."

Eaton hesitated a little but gritted his teeth and expressed his opinion.

"Honestly, Bob, you seemed more like a salesman than a scientist."

So, there it was, out in the open.

"Goddammit!" Heath roared at his resident. Then he turned and stormed out of the office. He slammed the door after him, so that its frosted glass gave way and shattered onto the floor. Eaton sat, frozen and in shock, in his chair. A few days before, he had driven Heinz Lehmann to the airport, and the old man had advised him to stay at Tulane.

"I'm sorry this is not working out for Bob, but he's a good man. Too impatient for research but an excellent clinician."

Now the shaken resident crept in to Irene and asked what he should do. Was he going to be fired, and if he was, where should he go? She calmed him down with the easy authority she mobilized with younger staff.

"Bob needed to hear the truth, and you are one of the few around here to give it to him. Keep it up."

CHAPTER 10

The Machine in the Mind

Arnold Mandell's eye-opening manuscript with the inside story of taraxein had materialized out of the blue, but tracking down its author took some work. There were plenty of e-mail addresses on the Internet, but none of them worked when it came down to it, and none of my contacts from the old Tulane group had any idea where Heath's former student had ended up. The trail petered out until a query in a chat forum for historians of psychiatry paid off. Mandell wrote from his home in Southern California that he would be happy to talk about Robert Heath. If I happened to be in the neighborhood I would be welcome to visit.

"But please keep in mind that I'm over eighty. I don't function until after eleven in the morning."

When I eventually showed up a at his La Jolla condo, it was a little past noon, and Mandell opened the door looking like a friendly troll that has crawled out of his cave: short, so stooped that a pair of suspenders had to hold up his pants, and sporting long, brushed-back bushy hair and a pair of large dark glasses to shield his degenerating maculae. The studio apartment was tiny, with a giant bed taking up most of the space, but it had a spectacular view of the ocean. We contemplated the water for a while and then sat down. Me on a kitchen chair and Mandell on the bed, propped up by pillows and rubbing a hand over his forehead every so often. His eyes were closed, for the most part, and his voice was corroded, but an intense mental energy nevertheless radiated from him. Arnold Mandell was a man on fire.

In 1969, he established psychiatry at the University of California–San Diego, and became the youngest doctor ever to occupy such a prestigious chair. A few years later, however, he was eased out after a scandal. As the house psychiatrist for the university's football team, Mandell had prescribed too many interesting—and illegal—medications to the players. The whole thing was described in a roman à clef called *The Nightmare Season*, which was published in 1976. Later, he wrote several books and held parallel professorships in San Diego and at other universities. Moreover, at eighty-two he was still the head of research at the privately run Cielo Institute and actively pursuing grants from the NIH.

"People always say—and it's even in *Wikipedia*—that my department in San Diego was the first in the country that was biologically oriented. But Tulane and Bob Heath had been since 1949," Mandell said to me from the bed. From his yellowed manuscript, which was in my bag, I already knew about his youthful fascination with Heath, but now Mandell told me that, even after so long, he himself never got over his time in New Orleans.

"I'm still turning over experiences from that time." Then he coughed, grunted loudly, and looked straight ahead. "It's as if Bob's spirit continues to haunt me."

As a twenty-year-old, the young man from California arrived at Tulane to become a doctor. With his grades from Stanford, he could have studied medicine at any university in the United States but decided against such top institutions as Harvard, Yale, and Princeton, because of Robert Heath. The rumors about a man who against all conventions went after the biology of the psyche had inflamed this precocious boy, and he joined the ranks. He was received with open arms. So, even as a student, he received full access to the laboratories and quickly began to move about freely as an observer among the department's researchers.

"Quite unusual access," he acknowledged. From others, I knew that Robert Heath always described Mandell as his most gifted student ever, one he took under his wing. He gave the young man a little lab to use at his own leisure, and the two met regularly behind closed doors to discuss science, ideas, and theories. But when I read Mandell's own narrative about his years with Heath, it exuded a fascination that I sensed had more to do with the man than his science.

"I'm absolutely heterosexual," said Mandell, raising a hand in solemn oath. "But I admit that there was an attraction to Bob's *intellect* that was almost erotic."

I asked about a passage in the manuscript of his novel. Arnold Mandell's alter ego, Aron, was about to leave the university and wanted to discuss things with his boss. This was toward the end of the book, after a fateful conference at which the critics with sharpened knives effectively halted financial support for the manuscript's Dr. White—that is, Robert Heath. White described himself to his young proselyte as "a mutant," a threat to the status quo that science so desperately maintained.

Dr. White compared himself resolutely with one of history's great figures, the British physician Edward Jenner, who in the 1700s developed the world's first vaccine. Jenner defied convention and prejudice when he infected his gardener's eight-year-old son with the deadly smallpox virus after having first immunized him with cowpox pus. The boy survived, and the method spread, but before doctors learned to control what they were dealing with, people died in droves along the way.

"Just like Jenner, I am a victim of living in a social and scientific transition period," said White. "I have accepted that."

The book recapitulated an extended discussion about the inherent conservatism of science and the built-in human need to explore and discover new things. Arnold Mandell confessed that he allowed his "White" to formulate himself more philosophically than reality's Bob Heath did.

"But I have no doubt that Bob felt he was chosen. He was simply put on Earth to do what he did, even if the world might not be ready for it."

But then there was the story about Matt Cohen and the big con that was never revealed. Arnold Mandell later found out about the real events from a biochemist named Grant Slater. Cohen had worked for a time as an assistant in Slater's lab at UCLA before he went to New Orleans. Cohen had introduced himself as a PhD to Robert Heath, and during his time at

Tulane, he flirted a bit with Mandell's sister Fanny, but she dumped him because he scared her. Cohen might turn up for a pleasant evening out and tell her how the car's trunk was filled with loaded rifles. He was nervous about being assassinated, he would explain.

Fanny also told her brother that, after the historic Macy conference, Cohen fled back to his hometown and his past in New York, where he had Mafia contacts. Only several years later did he pop up again on the radar screen when Mandell happened to read a notice in the paper about a dustup concerning some casino ships off Miami. It related that Matt Cohen had been among those killed.

"A crazy story," said the old man, shaking his head. As a young researcher Mandell had first been close to Robert Heath and later to Seymour Kety. He had great respect for both, he told me, but for very different reasons.

"Kety was meticulous and saw himself as the conscience of psychiatric research at the time. And God knows a lot of sloppy research was going on in the field," Mandell said. "But he had contempt for Heath and used him as exhibit A for what was wrong with psychiatry. I believe he wanted to make an example of him and latched onto taraxein."

Righting himself on the bed Mandell began to massage the wrinkles in his forehead. He never understood why Heath had garnered as much hostility as he did with all the "half-assed" science going on in many places. But had Heath not set off Kety early on but just stayed with his brain stimulation work, he said, Heath might have fared much better.

For me, the big question was why Heath had decided to go to the pivotal Macy conference with a compromised biochemist in tow—a mobster who had taught himself some protein chemistry on the side. That seemed reckless, even arrogant bordering on dishonest. He should have postponed, replicated everything, and taken new data.

Mandell nodded slowly. From time to time, he said he had asked himself whether there was something slightly psychopathic about it, an ability to repress completely what others might say and believe about something. But that couldn't be true, because he found Robert Heath to be a man of empathy and human understanding. Also, Mandell had to admit that Byron Leach, who was close to the clinical side of schizophrenia and whom he knew to be "the most straight and honest guy," was always convinced that the taraxein observations represented something real.

"Heath had a high purpose and high resilience," Mandell said, echoing something another of Heath's students, James Eaton, had told me. That Robert Heath was a believer and that believers are the ones who move science ahead while the bookkeepers fill in the gaps.

"What strikes me today, is how Heath was a forerunner in several areas of psychiatry that are now coming up," continued Mandell. He remembered his mentor arguing in the 1950s that there must be connections between the cerebellum and frontal areas of the cerebral cortex. He couldn't prove it at the time but found those connections two decades later and used them to successfully treat a handful of schizophrenic patients with a cerebellar pacemaker. No one paid any attention at the time, but now the idea of the cerebellum's involvement in schizophrenia is enjoying its day in the sun. In 2014, Nancy Andreasen and her colleagues at the University of Iowa even proposed stimulation of the cerebellum as a treatment—presenting the idea as a "novel hypothesis."

"As for taraxein," Mandell said, looking directly at me with an almost sad smile, "the connection between schizophrenia and the immune system has become a hot topic. Think if Bob turns out to be right."

Mandell's remark prompted a vague memory and made me to go back and search my archive of research material. From one of the messy piles I pulled a paper titled "The Search for an Endogenous Schizogen: The Strange Case of Taraxein." It was from 2011, and in it Alan Baumeister, the psychology professor I had met in Baton Rouge, went through Heath's experiments and critiqued the evidence for the sensational protein brick by brick. But as he wrote at the end: "Even if it is improbable, it is possible that Heath made an important but not yet recognized discovery." And he went on: "Taraxein may exist but only in a subgroup of patients under very special circumstances."

I dug into the immunological research on schizophrenia and saw that Arnold Mandell was right. It *is* a field on the way up. "The return of immunopsychiatry" was the focus of a special issue of the journal the *Lancet Psychiatry* in 2015. As for schizophrenia, the literature is ripe with reports of patients who exhibit chronic inflammation, overproduction of certain cytokines, and whose general immune response is compromised. The findings have prompted researchers to experiment with adding anti-inflammatory drugs to the regular antipsychotic medications. Good old aspirin has been tried, as have other NSAIDs, as well as the growth factor EPO and the antibiotic minocycline. Several recent reviews conclude that results are preliminary but promising, especially for improving cognitive symptoms. Meanwhile, one can find ongoing clinical trials of more specific drugs like Sylvant, or siltuximab, which targets the cytokine IL-6.

I discovered that what brought this interest to Parnassus was a Danish register study from 1999. At that time, Preben Bo Mortensen from the University of Aarhus was poring over a store of data from almost two million Danes born between 1935 and 1978. He found that the risk of developing schizophrenia over the course of a lifetime differed depending on time of birth. It was highest for people born in February and March, while you had the lowest risk if you came into the world in August and September. It sounded crazy. But Mortensen's hypothesis was that it had to do with the extent to which the pregnant woman had been subject to infections from, among other things, the influenza virus. The invasion of microbes might have left traces in the embryo's immune system, which later came to play a role in the outbreak of schizophrenia.

In 2009, genetic research pointed in the direction of the immune system. Three research teams almost simultaneously discovered a very strong link between mutations in the so-called MHC genes and an increased risk of schizophrenia. MHC (which stands for major histocompatibility complex) proteins are some of the key proteins in the immune system. They sit on the surface of our cells where they interact with pieces of wayward viruses and bacteria and, thus, combat infections.

In addition, there have been observations suggesting autoimmunity plays a role in schizophrenia. When a person's immune system attacks itself, a range of diseases can develop (including rheumatoid arthritis). In 2013, Mortensen and colleagues pointed to a striking link in the Danish population registers. The data showed that people with schizophrenia suffered almost three times as frequently as the population generally from a series of autoimmune diseases, including lupus, psoriasis, and type I diabetes. Moreover, the autoimmune ailment reported itself before the schizophrenia broke out, which could indicate a causal connection.

In that same year, one could read in the *Journal of the American Medical Association* about the discovery of special antibodies in acute schizophrenics—antibodies against a protein in the brain. It concerned the so-called NMDA receptor, which plays a central role in higher brain functions, such as learning and memory, and is activated by the neurotransmitter glutamate. German researchers from the University of Magdeburg tested blood samples from 459 patients with schizophrenia, depression, or borderline personality disorder and compared them with 230 normal matched control subjects. Only in schizophrenics did they find an increase in antibodies against the NMDA receptor, and the increase did not apply to antibodies against other tested receptors. The Germans were not the only ones to investigate the matter. In 2014, a meta-analysis in *Schizophrenia Research* of nine studies concluded that the different data sets reliably pointed in the same direction.

In 1967, nearly fifty years before, Robert Heath and his colleague Iris Krupp found autoantibodies connected to brain tissue in schizophrenics but not in control subjects. At the time the observations did not attract much attention. In contrast, the association between NMDA antibodies and psychosis has been called "One of the hottest, even inflammatory topics in neuropsychiatry today." It precipitates disputes about how and where to measure the antibodies—in serum or the cerebrospinal fluid bathing the spinal cord—and arguments about whether the antibodies might be causative or rather just a marker for a general neurological vulnerability. Discussions that are eerily reminiscent of the long-forgotten taraxein debate.

This immune-system research sent me to my bookshelf to get down a copy of the old *Studies in Schizophrenia* from 1954. Wasn't there something about arthritic patients getting better from electrode stimulation? Arthritis has long been the most well-known autoimmune ailment.

At his 1952 Tulane symposium on schizophrenia, Heath talked about a single patient with severe rheumatoid arthritis and how he reacted surprisingly positively to electrode stimulation. The text mentioned that the man's acute pain disappeared and was moderated for an extended period, and there were numbers for how his immune system reacted. The number of white blood cells—the storm troopers of the immune system—fell abruptly to one-fifth of what they were, and the steroids in his body were affected. The patient's cholesterol numbers fell similarly and, parallel to that, an unidentified but new steroid suddenly turned up in his blood. They had no way of finding out what sort of a molecule it was, but the change itself showed that the immune system was reacting.

I dug out the old films, searched through them, and found even more footage of arthritis patients. The first was Mr. Needham. I could see from the dates that he must have been the man mentioned at the symposium. For seven years, he had been crippled by rheumatoid arthritis. He was thin with wasted muscles, and you could watch him scuttle over to a chair and sit down. He was unable to stretch out his legs, and he could not step up on a low box without help. Even from a distance of seventy years, the scene was painful to observe.

But then there was a cut to the stimulation as it was happening, and, presto, we saw a completely different man. Mr. Needham was lying on his white bed and stretching both legs at a quick tempo while he muttered some strange, fascinated sounds and called first on God, then the doctor.

"Oh, Doctor . . . oh, it feels so good."

A week later, the two met again in front of the camera. Mr. Needham's joints were no longer red and swollen. The inflammation had abated, and the pain was pretty much gone, he said. Only the right knee, whose bone tissue the disease had irreversibly broken down, still gave him problems.

In his presentation at the Tulane symposium, Heath himself called his findings interim but provocative: "If our hypothesis is correct, further research may show that many of the alterations in the internal milieu of an individual which make him susceptible to somatic diseases and which can influence the progress of disease are influenced by the thought."

In his convoluted way, Heath was saying that our thoughts, or our psyche more broadly, could be influencing our physiology and, through that, whether we get physically ill or how an illness proceeds. He stressed that he wanted to pursue this idea, and in the films I found more cases like Needham's from the following years. A slightly later patient was Jackie, a fifteen-year-old black girl who was terribly slim and almost incapable of moving. She was carried in and put on a bed, where she whined about the touch of the white lab coats. She could neither stretch nor bend her legs, and both knees and elbows were swollen and warm from inflammation. But an hour later, the stimulation in the septum had apparently done its work. She could freely move her wrists, ankles, and knees and did not have pain. In the next clip, it was 1962, and Jackie was again visiting Tulane, now as a married woman and mother of a small child. She was still thin as a reed and stiff in certain joints but stated that she had not had pain since her treatment eight years earlier.

What happened in these cases? There was no control group, and it may have been a placebo effect. Moreover, immunological responses are notorious for swinging up and down in intensity—they can also vary with age. Whether the stimulation in the septum was effective, no one could say on the basis of the pictures. On the other hand, there can be no doubt that this *approach* was far ahead of its time. Heath's writings about how thought may affect physiology fit perfectly into the modern research field of psychoneuroimmunology, or PNI. PNI was coined as a term by the psychologist-immunologist team Robert Ader and Nicholas Cohen in 1975, but the field has emerged after the 1980s. Before then, it was considered unthinkable that there might be a direct communication between neurons and the immune system, but such interactions are now showing up at different biological levels. Findings seem mainly to center on the somewhat cloudy concept that we call stress. If stress hormones such as cortisol and glycocorticoids are chronically increased, they can alter the brain's signaling via neurotransmitters. Various proinflammatory molecules produced by immune cells can act on neurons and affect their growth and function. On the other hand, endorphins produced in the brain can interact with certain cells in the immune system.

These intimate connections between nervous tissue and immune system are also a focal point within a budding discipline dubbed bioelectronic medicine. The vision is to replace chemical medications by instead modulating the biology of the body with properly directed electronic signals. In the pharmaceutical industry, a number of major players have been establishing their own bioelectronics research programs, and the National Institutes of Health, which has the money, announced in 2014 that over the coming years they would grant somewhere in the area of $1.7 billion in research support.

Current clinical experiments stimulate the nervous system in order to treat chronic rheumatoid arthritis. The originator of the technique was Kevin Tracey, who heads New York's Feinstein Institute for Medical Research, and he uses the so-called vagus nerve as his entry port. This large, thick cranial nerve emanates from the medulla oblongata, runs along the carotid artery, and sends offshoots to all the body's major organs on its way to the stomach. Through its connection to the spleen, it affects the immune system, and Tracey discovered in the late 1990s that electrical stimulation of the vagus nerve can suppress the production of an important immune molecule: the tumor necrosis factor, or TNF, which creates inflammation and is overproduced in immune diseases such as rheumatoid arthritis.

For some years, Tracey studied the vagus connection in detail on rats, and after a pilot trial on people, he established a company, SetPoint, and started commercializing his findings. SetPoint developed a small stimulator that could be implanted around the vagus nerve, and in 2011 they began clinical testing of vagus stimulation for the treatment of rheumatoid arthritis in patients. And this has caused quite a stir. Of the eighteen people severely affected by arthritis who ended up participating, twelve experienced a positive long-term effect, which sounds a lot like what Robert Heath reported in Jackie, Mr. Needham, and others. The inflammation in the joints diminished, and pain abated significantly.

Random? Possibly not. At any rate, various recent studies investigating areas of the brain that have a connection to the vagus nerve have mentioned something familiar—namely, the tiny nucleus accumbens, which is a part of Heath's septum. This raises the question: Could Heath in his day have also stimulated the vagus nerve when he was stimulating the accumbens, and thus gained access to the immune system? Is that how he reduced arthritic pain?

Robert Heath's original idea about treating schizophrenics with electrodes in the pleasure center is still alive.

Psychiatrist Holly Moore and her colleagues at Columbia University recently suggested that you could do deep brain stimulation in either the nucleus accumbens or the hippocampus to treat schizophrenia. In her publication Moore mentioned quite briefly that Robert Heath was the first to stimulate schizophrenics in the nucleus accumbens but did not delve more deeply into his work. And when she explains her own rationale, she doesn't reference firing up the pleasure center but modulating the production of the transmitter dopamine. Recent scan studies have shown that schizophrenia goes together with an increased activity in the hippocampus and that this activity presumably creates an excess of dopamine in other areas of the brain in the pleasure system.

Moore has a hypothesis that stabilization and moderation of neural activity with high-frequency stimulation in the hippocampus or the nucleus accumbens would treat what are called the *positive* symptoms of people suffering from the illness, which is to say hallucinations and delusions. The authors concluded: "Thus, we suggest a direction for further experimentation aimed at developing neurosurgical therapeutic approaches for this devastating disease."

It was like reading Heath's accounts about how acutely psychotic patients were calmed by stimulation in the septum. Both Heath and Moore, half a century later, found that their subjects' hallucinations were characterized by increased activity in the hippocampus.

But it got even more interesting when I began to wander around the large American database on clinical trials. That I would find nothing seemed very likely. When I typed in *deep brain stimulation* and *schizophrenia*, there was a single hit. In

Toronto, psychiatrist Jeff Daskalakis reported an experiment that stimulated areas of the nucleus accumbens and the ventral tegmentum to counteract the *negative* symptoms of schizophrenics. That is, apathy and anhedonia!

Déjà vu. It was like hearing an echo of Robert Heath.

I caught the Canadian psychiatrist on his cell phone as he was on his way to work. Daskalakis told me that for two years, he and his colleagues behind the experiment could not recruit so much as a single patient. The problem was, he complained through the noise on the line, that the patients he was trying to reach had no motivation to seek help; their affliction robbed them of such initiative. I could sense that this hurt the young surgeon, who was convinced his proposal could help. As he said at one point, "It *is* just so logical to stimulate their reward system."

Over a few minutes, I sketched Heath's story, and I told him about the many publications and descriptions that existed. Daskalakis was completely silent on the other end of the line, and when he finally answered, it was with disbelief in his voice.

"I have never heard about any of this. And in the 1950s. That sounds . . . wild."

There is indeed something wild about the story of Robert Heath. After all of my reading, all of my visits, and all my transatlantic conversations, I am left with a feeling of astonishment. It is like going into an old attic and opening a dusty cardboard box that turns out to be full of sparkling knickknacks. Many months ago, I started to examine a bizarre experiment with a gruesome fascination and curiosity about the shadowy figure who was behind it. I had a vague sense of a story and just had to know what really went on and who this man was.

What did I learn?

My original impression of perverse research and white-coated callousness, of Robert Heath as a despicable figure who, without any consideration for anything other than his own ambition, experimented on the weak no longer holds up. It is not fair that his intellectual legacy is pretty much based on his experiment on B-19.

The criticism of the old research and even the research itself, the colleagues who participated in experiments or observed them from a distance, the one still-living patient and his nostalgia, the forbidden films of long-deceased patients whom posterity describes as victims but who, as far as I can see, look more like partners in collaboration with their therapist—this evidence does not paint the portrait of a monster.

My picture is not exhaustive or an expression of the final truth. Despite everything, I have only been able to speak to a few of those who knew Robert Heath or were under his care. And he is still a mysterious, interesting, and alluring figure. But as Arnold Mandell explained to me: so he remains even for people who were close to him.

Something the brother of Heath's secretary Irene, Colby Dempesy, said to me at one point resonates: "The Heath story is a sad one, but it is not negative." At the time, I thought the phrasing was strange—what did that mean? But now I think I know. Robert Heath's head-on collision with the zeitgeist and the way his own personality stood in his way *was* sad. He was an original, willing to stick his neck out. But he was too headstrong to garner acceptance for his ideas and too exclusive to participate in the scientific community. Finally, the arrogance, which was perhaps the dark side of being unconventional, paved the way for making bad decisions.

He began in the stratosphere with an unheard-of privileged position and some promising ideas. He stayed at the top of his profession for a while but then took a nosedive as his reputation outside the microcosm and protective workplace of Tulane University fell. In its way, the end says it all.

In 1980, Robert Heath turned sixty-five, which was Tulane's official age for retirement, and he became an emeritus professor. He was freed from all administrative duties but continued his own research along with his faithful squires, Irene, and the technicians Herb Daigle and Charlie Fontana. They kept working with, among other things, the cerebellar pacemaker. Yet even though a surgeon in Alabama tested it a few times, it never garnered any great interest from the outside, and a couple of years later, the operations stopped. The FDA classified the procedure as purely experimental, so insurance companies would not pay for it.

Robert Heath himself was having more and more problems with his heart, and in 1986 he finally had to stop working and leave his remaining patients to his former partner Donald Richardson. With his wife, Eleanor, he moved from New Orleans to Florida, where he played golf and worked on his monograph. *Exploring the Mind-Brain Relationship* was the book that was going to epitomize his work and that he imagined would make clear to everyone once and for all the purpose of his decades-long project.

I'm guessing that he saw the book as an opportunity to avoid oblivion and get out from under the long shadows of the 1970s. A final chance to achieve the recognition he believed he deserved. For years, he puttered around with the text and rewrote it again and again. But when it came down to it, he could not find a publisher to print the book. Finally, he had to swallow his pride and pay for the production of a few hundred copies from an obscure printing house in Baton Rouge. The books were printed in 1996, just three years before his heart gave out. He sent a handful to his old colleagues and received nice thank-you letters on embossed paper.

"When we initiated the program, we were young and vigorous and expected to have many answers within a short time," Heath wrote in his introduction, with an eighty-one-year-old's ironic distance to his younger self. He made it clear that the group was far from getting all the answers they had hoped for, and the book ended in resignation: "If our findings inspire other investigations or contribute to advances in years to come, they will have served their purpose."

But the work at Tulane remained in obscurity, and no contemporaries were inspired by it. Nevertheless, it is in relation to the work that I understand Dempesy's remark that the Heath story is not negative. Because the *project itself* was not objectionable, and his line of investigation has reappeared, riding a wave of great enthusiasm. Robert Heath was not a mad scientist; he was a pioneer.

But was he a pioneer by accident like a blind chicken that still manages to find the corn? Do his findings simply by chance happen to point in directions that now seem to be very interesting?

When I consider the published studies together with the testimony of old colleagues about a sharp thinker with a deep, broad knowledge of literature, I do not believe it can be said that Robert Heath was working in the dark. Rather, it seems as if he had a vision of something of which he could not clearly see the contours—quite simply because science had not yet reached far enough and the tools were still primitive. Heath and his generation were working with homemade electrodes and old-fashioned EEG devices, and they knew vanishingly little about the chemistry that unfolds inside the skull. Half a century ago, data was much harder to come by and to process, and typically comprised the limiting factor in experimentation. Formulating a plausible hypothesis and working with what you had in order to test it was crucial. We who look back today live in a completely different world. For one, much of science is now data driven. It centers on letting supercomputers process massive amounts of information to identify patterns that might then serve up unknown links and interconnections. Genetics is a prime example, with still larger sequencing projects being launched left and right. As for neuroscience, we have grown accustomed to its practitioners using nifty scanners to peer into the living brain and to their having an overview of the abundance of the organ's chemical signals.

It is precisely here that the story of Robert Heath and his struggle with deep brain stimulation opens up. Heath points beyond himself and grows into something bigger because, in parallel to what is going on with deep brain stimulation today, his work becomes a mirror for us.

"History is repeating itself. The only difference from that time and now is that the tools are more sophisticated. Otherwise, it is the same conduct as before—also the same regrettable conduct."

I heard this remark at a get-together for neurosurgeons in Maastricht. The speaker was Marwan Hariz, a seasoned, old neurosurgeon at University College London and the field's self-appointed historical memory. It is Hariz who stands up at scientific conferences to talk about ethics and the sins of the past in the hope that his contemporaries will learn the lesson. When I first heard him, I thought the conclusion sounded a bit like dramatic exaggeration—modern scientists, after all, have a completely different and more thoughtful approach to their work than the headstrong medicine men of the 1950s. Perhaps the curly haired surgeon with a Lebanese accent is onto something.

Psychiatry is where deep brain stimulation began its history. Just like back then, we are still talking about a strikingly small group of patients who are followed far too carelessly in the literature. And behind the various experiments, there are still enterprising individual surgeons, neurologists, and psychiatrists, each of them developing and marketing their special methods and their special areas of the brain—and becoming well known and celebrated for it. As Hariz puts it, "Ego and personal prestige are some of the most powerful engines in the field."

A new driving force has also appeared: lots of money. The money comes in because the companies that are producing the equipment itself are pressing health authorities to approve new treatments and, therefore, pursue clinical trials too early. This is happening, as both Helen Mayberg and Thomas Schläpfer told me, with electrode stimulation on depressed patients. On the basis of data from a single group, major trials were launched at several hospitals, and they failed horribly. We may see the same thing with brain stimulation for combatting Alzheimer's. It was immediately proclaimed to be sensational when Andres Lozano operated on six patients, two of whom apparently had the degeneration of their brains halted and their memory become a little better. These two patients were only followed for a year after the operation, and no one has since heard anything about how they are doing. Nevertheless, the Medtronic company has thrown itself into a trial that is taking place at several hospitals in the United States and Germany. The heavy commercialization of all things medical has had its costs and benefits.

The media typically functions as a squad of cheerleaders. At the University of Tasmania, ethicist Frédéric Gilbert has studied the coverage of deep brain stimulation, and in *Frontiers of Integrative Neuroscience*, he points out how the media attention is extremely positive, thoroughly uncritical of the researchers' claims, and part of creating a dangerous hype. Journalists who do not ask why the experiments are so limited and scattered and why patients are not followed up on help create unrealistically high expectations among readers. It is disturbingly reminiscent of the original coverage of lobotomies. In the 1930s and 1940s, there was such praise that it created a demand for the procedure among the public—families who wanted their insane relatives treated. Not until around 1960, after the introduction of the first antipsychotic medications, did the coverage begin to mention the lack of efficacy and harmful side effects of the procedure, and the lobotomy quickly fell out of favor.

Could there be a backlash against brain stimulation like the one that hit lobotomy and psychosurgery generally? In principle, yes, and Marwan Hariz continues to warn his colleagues about it. I think it is likely that stimulation will develop from an experimental to a routine treatment. Not just because the techniques and the tools *are* better than before but because the spirit of the times is completely different.

Our sense of who we are has changed—the self is no longer an incomprehensible and intangible phenomenon but is conceived as a state in the brain. Something strange has also happened to our relationship to psychiatry itself. It has become common property. From being an obscure, marginal field in the world of medicine, psychiatry has moved into our collective consciousness, where it looms larger and larger. Once again, we can look to its presence in the media, where there is an almost endless debate about mental illnesses, their treatment, and their significance. A generation ago, ordinary people wouldn't even talk about such things, whereas today we jabber happily away about our psychological defects.

You might say that there has been a psychiatrization of our

view of human beings. We have reached a point where, more than anything else, psychiatry is the prism through which we think about and understand ourselves. Character traits, personality, and overall behavior are held up automatically against the diagnoses we think we know so well. We observe our neighbor's difficult, noisy child and feel convinced that he has a mild case of ADHD. Or we debate with our work colleagues at lunch about how high the CEO would score on the psychopathy scale while we secretly think that our introverted colleague at the end of the table probably fits somewhere on the autism spectrum with more than a touch of Asperger's syndrome. Various ailments become a question of a spectrum rather than a sharp distinction between sickness and health and, thus, more behaviors are now seen as diseased or, at least, inexpedient. And in turn, this provides even more leeway to what we can permit ourselves to correct or optimize.

Take something like porn addiction. People no longer talk about a lack of backbone if people are glued to the screen for hours looking at naked pictures and sex films—no, it's a disease, and of course, a treatment is offered. Naturally, there is old-fashioned psychology and talk therapy, but it is not always effective. So, why not proceed more scientifically and mechanistically? This is what neuroresearcher Nicole Prause from UCLA does. She has studied what happens in the brains of people who cannot stay away from porn, and she has observed which areas show abnormal activity. Among other things, she now investigates how deep brain stimulation can be used to relieve the problem by dialing down the desire for sex. As she said to the *Huffington Post* in 2015, "This issue is important so that therapists can be held accountable to their patients for providing treatments that have been shown by science to help."

Prause is right. It is easy to imagine that deep brain stimulation will expand its territory in the future because we are dealing with a field undergoing sweeping technological development. The considerable investment by manufacturers and DARPA's large grants and equally large media hype have led to tremendous refinement and advancement in electrode systems, which can be directed at ever fewer cells and perform more and more specialized tasks.

This progress and these visions create a force field and an attraction. In addition, there is a special dynamic that comes into play: When you *can* do something, an unforeseen need to *do* it often arises. We have seen this very clearly with various psychopharmaceuticals. When SSRI medications (such as Prozac and Zoloft) first came on the market, they were approved for the treatment of anxiety symptoms and depression in adults. The medications quickly spread to other patient groups—particularly children and young people—and a whole ream of other conditions. Whether the problem is compulsive thinking or shyness, it can be treated with a certain success by SSRIs.

Medications such as Ritalin and Adderall are amphetamine-like stimulants, both of which were developed and approved for the treatment of ADHD symptoms. But now they are just as well known for their role as intellectual performance enhancers or cognitive doping. If you take them for a brief time, the medications increase your concentration and ability to work, and it is well known that many students use them when they have to write papers or take exams. And students are not the only ones. A few years ago, I met an older British colleague who told me that she always wrote her books on Ritalin. “It is *so* much easier and quicker,” she said.

Improvements in electrical equipment are also taking off. Here, too, the vanguard is not the professionals—the system—but ordinary consumers: people who, with the tools at hand, try to boost their concentration or memory. There is a minor do-it-yourself movement of people who use small headsets and 9-volt batteries to send electricity through their cranium to hit the outer part of the cerebral cortex. First movers, like the Williamses of Atlanta, told their story to *Wired* in 2014. As an engineer, Brent Williams had no trouble cobbling together a device from parts procured at Radio Shack, and while he uses it to get a sort of “runner’s high,” his wife, Madge, finds it helps her better remember scripture.

Transcranial direct current stimulation, as it is called, has gained a certain following, with happy media accounts about how the method apparently increases learning ability in American military personnel. And for those who cannot themselves construct a device, the company Foc.us sells small, streamlined apparatuses over the Internet, marketed in particular to the gamer crowd. *Hyper-charge yourself* is the offer to computer gamers who want an extra *edge* in front of the screen. In this way, people get around the requirements of the health authorities for the approval of actual medical technology.

It is easy to laugh at these do-it-yourself folks, and, of course, it is a long way from a homemade device with a 9-volt battery to highly specialized brain surgery with advanced electrodes directly connected to nerve tissue. But development is nudged by the fact that we live in an electronic age. We are all thoroughly used to electronics and, particularly, to being completely dependent on electronics. We use our phones and computers even though we know that they are under surveillance by various security agencies. You could say we already live life as cyborgs—it's just that the electronics are still, for the most part, outside the body. For how long will that be the case?

As recently as 2006, the American neurosurgeon Katrina Firlik and author of the book *Another Day in the Frontal Lobe* said to the *Evening Standard* that "one day soon, we'll all have brain-lifts, just as we have Botox now." And in 2011, a survey of her colleagues in the American Society for Stereotactic and Functional Neurosurgery revealed that half of them saw no ethical problems in using deep brain stimulation to improve mental capacities.

And why should they? After all, everyone gets older, and everyone wants, above all, to retain their full mental capacity. Both society and the individual have a great fear of Alzheimer's and dementia generally. Modern life does not leave room for retirement. It demands that we keep ourselves in top form and work until we collapse and that we continue to adapt. Add that to another argument that is often heard—namely, that people are equipped with a Stone Age brain that is ill fitted to our high-tech reality. Today, the natural reaction is to conceive of technological modification and intervention as a solution. A Stone Age brain must, of course, be upgraded, so it can keep up. The tempo cannot be slowed, or progress will be hindered, right?

It would seem that there is pretty much nothing we would not at least consider playing around with. In 2015, under the headline "Morality 2.0," *New Scientist* asked whether it might not simply be necessary to manipulate human morality in order to save the world. Our moral intuition and capacity for empathy, the argument goes, are shaped by our evolutionary past and adapted for a life in small groups in which people are close to one another. Therefore, we are wired to relate to things that are very concrete to us as individuals but not very good when it comes to tackling more general and complex problems—in other words, exactly the kind of global challenges that modern humanity faces: the climate crisis, the destruction of the oceans, mass migration. Various researchers suggested to *New Scientist* ways to train people's moral thinking, so our reflex-like intuition takes a back seat to more conscious reflection. But others believe that this sort of thing is not enough. In their book *Unfit for the Future—The Need for Moral Enhancement*, two philosophers, Ingmar Persson from the University of Gothenburg and Julian Savulescu from Oxford, argue for the use of biomedical drugs. These two thoughtful gentlemen have no doubt that the brain must be manipulated if things are going to work.

Once again, I hear the mesmerizing echo of Robert Heath.

In 1985, Heath wrote a leading article in *Biological Psychiatry* in which he asked the question: Should we see the human brain as a tool for progress, or catastrophe? He himself believed there were clear signs of the latter. These were thoughts that he must have had from the beginning, because in Arnold Mandell's tale of Tulane in 1950s, he had his Dr. White say: "Survival will one day simply require central nervous system engineering." And in Heath's own monograph, *Exploring the Mind-Brain Relationship*, one of the last sentences dealt with the same thing: "If one day a common moral code evolved, and biologic methods could be applied so as to imprint firmly the memories of that code, it might be possible for man to live in harmony with his fellow-man, as well as with other species."

Epilogue

The two men were sitting, partially turned toward each other, with a low table between them. The location and the furniture were an indeterminate gray, but someone quickly brought out some yucca plants to add a little green to the background. One of the men cleared his throat.

"Good evening," he solemnly intoned, looking into the camera that was recording the scene. "And welcome to this conversation with Robert Galbraith Heath, who was a professor and head of neurology and psychiatry here at Tulane between 1949 and 1980."

Wallace Tomlinson had been looking forward to this interview. As a young psychiatrist, he had been a resident under Robert Heath and had always seen the older man as a mentor. Now he had persuaded Tulane to begin a collection of oral histories from the figures who had meant something to the department over time. This was 1987, seven years since Heath retired, and Tomlinson himself had become a rotund man with heavy glasses and just as heavy diction. He was sweating in his gray polyester suit and thought how characteristic it was that the seventy-six-year-old Heath—as always—had a cool, almost stylish charisma. Heath wore a blinding white shirt, gray pants, and a black jacket with an impeccable bow tie. His hair was completely white but still lush, and his eyes were clear.

"Let us begin with what got you interested in medicine, Dr. Heath."

The old man laughed softly.

"To be honest, I was pressured into it by my father, who himself was a doctor and had a practice in Pennsylvania. But I came to my specialty as a neurologist later through my wife's father. He was a big influence."

And then there was psychiatry, which the army had thrown him headlong into during the war and which he stuck with. As planned, Tomlinson guided the discussion through Heath's curriculum vitae and made sure they touched on various honors and lost themselves in memories of great figures of the past.

"What about Sandor Rado, who was a mentor for you? Tell us about him."

Tomlinson noticed how the tone changed at once when Heath spoke about his old teacher.

"He was a very inspiring and gifted person, very creative, and with a more fundamental understanding of human behavior than anyone else at that time. Or later, if you ask me."

Even Heath's voice had suddenly changed. It was more insistent, thought Tomlinson, and more attentive. He could sense that something was on the way. Something was on the well-dressed, white-haired gentleman's mind.

"It's too bad," Heath said slowly, "that Rado did not receive the recognition he deserved."

Wallace Tomlinson blinked behind his heavy glasses as if realizing that Heath was no longer talking about Rado at all, but about himself. This was about his legacy. Wallace Tomlinson was deeply interested in history, and he often thought that, at some point, he needed to write something about his boss and mentor. He had to try to describe the Robert Heath he knew and appreciated, for posterity, and to explain why he was such a controversial figure. But he was pulled back to the moment as Heath expanded on his thought.

"Rado was controversial, and he had many people around him who embraced the ideas for which he was criticized, but when they later turned out to be correct, people began to forget where they came from. As a result, many of the new and creative approaches he introduced continued to be developed but were gradually decoupled from him. Rado is pretty much *forgotten*."

The old man kept a steady countenance but, at the same time, had a slightly agitated expression in his eyes. Then he went on.

"Of course, this is a phenomenon you often see when you're talking about gifted, creative individuals, and especially if they are not the smiling, pleasant, political type. Rado wasn't. Creative people often say what they mean and go against existing ideas."

Heath stopped. He had said what he had to say about the matter. Tomlinson wanted to engage with him and tell him that he understood. With the former disciple's concluding remark, it was as if a private code was passed between the two men.

"History often corrects itself as time goes by," said Tomlinson, "and people assume their rightful place in the sun. I feel this may very well happen for Sandor Rado."

SOURCES

I want to emphasize that this book is a work of nonfiction. Besides the written sources listed below, the historical chapters build on my meetings and extensive interviews with people who either worked directly with Robert Heath or have met him in some other capacity. I have only presented scenes, situations, and occurrences that I have found in the written source material or have had described to me by interviewees.

Interviews have been conducted over a long period of time and apart from many phone conversations across the Atlantic I have met with my sources in the following places. April 2013: Long Island. May 2013: Montreal. July 2013: St. Kitts. February 2014: Washington, DC; Philadelphia; Baton Rouge; New Orleans; Riverview, FL; and Atlanta. July 2014: Miami and New Orleans. September 2014: Maastricht. June 2015: Picayune and New Orleans. February 2017: Boston and La Jolla.

For general background and context, I consulted:

Delgado, José M. R. *Physical Control of the Mind: Toward a Psychocivilized Society.* New York [etc.]: Harper & Row, 1969.

Ferber, Sarah. *Bioethics in Historical Perspective.* London: Palgrave Macmillan, 2013.

Fradelos, Christina Kathryn. "The Last Desperate Cure: Electrical Brain Stimulation and Its Controversial Beginnings." A dissertation submitted to the faculty of the division of humanities in candidacy for the degree of doctor of philosophy. University of Chicago, 2008.

Heath, Robert Galbraith. *Exploring the Mind-Brain Relationship.* Baton Rouge, LA: Moran Printing, Inc., 1996.

——. *Studies in Schizophrenia: A Multidisciplinary Approach to Mind-Brain Relationship.* Cambridge: Harvard University Press, 1954.

Heath, Robert Galbraith, ed. *The Role of Pleasure in Behavior: A Symposium by 22 Authors*, xiv, 271 illus. Includes bibliographies. New York: Hoeber Medical Division, Harper & Row, 1964.

Hooper, Judith, Dick Teresi, and Isaac Asimov. *The Three-Pound Universe.* New York: Dell, 1986.

Mandell, Arnold J. *Psychosurgery—1954*, 1954.

Mohr, Clarence L, and Joseph E Gordon. *Tulane: The Emergence of a Modern University, 1945–1980.* Baton Rouge: Louisiana State University Press, 2001.

Salvaggio, John E. *New Orleans' Charity Hospital: A Story of Physicians, Politics, and Poverty.* Baton Rouge: Louisiana State University Press, 1992.

Valenstein, Elliot S. *Brain Control: A Critical Examination of Brain Stimulation and Psychosurgery.* New York: Wiley, 1973.

Winter, Arthur. *The Surgical Control of Behavior; a Symposium.* Springfield, IL: Thomas, 1971.

For detailed, specific research, I consulted:

Abbott, Alison. "Neuroscience: Opening up Brain Surgery." *Nature* (October 2009). doi:10.1038/461866a.

Abramson, Harold Alexander, and Josiah Macy. "Neuropharmacology: Transactions of the Fourth Conference: September 25, 26, and 27, 1959, Princeton, NJ" The Foundation, 1959.

Baumeister, Alan. "The Search for an Endogenous Schizogen: The Strange Case of Taraxein." *Journal of the History of the Neurosciences* 20, no. 2 (April 8, 2011): 106–22. doi:10.1080/0964704X.2010.487427.

———. "Serendipity and the Cerebral Localization of Pleasure." *Journal of the History of the Neurosciences* 15, no. 2 (July 1, 2006): 92–98. doi:10.1080/09647040500274879.

———. "The Tulane Electrical Brain Stimulation Program: A Historical Case Study in Medical Ethics." *Journal of the History of the Neurosciences* 9, no. 3 (December 1, 2000): 262–78. doi:10.1076/jhin.9.3.262.1787.

Behar, Michael. "Can the Nervous System Be Hacked?" *New York Times*, May 23, 2014.

Bishop, M. P., S. Thomas Elder, and Robert Galbraith Heath. "Intracranial Self-Stimulation in Man." *Science* 140, no. 3565 (April 1963): 394–96.

Bourzac, Katherine. "Neuroscience: Rewiring the Brain." *Nature* 522, no. 7557 (June 2015): S50-2. doi:10.1038/522S50a.

Canavero, Sergio. "Criminal Minds: Neuromodulation of the Psychopathic Brain." *Frontiers in Human Neuroscience* 8 (2014): 124. doi:10.3389/fnhum.2014.00124.

Carter, Cameron S., Edward T. Bullmore, and Paul Harrison. "Is There a Flame in the Brain in Psychosis?" *Biological Psychiatry* 75, no. 4 (February 2014): 258–59. doi:10.1016/j.biopsych.2013.10.023.

Catholic Online. "Paging Dr. Frankenstein: Agency to Research Brain Implants," November 4, 2013, http://www.catholic.org/news/health/story.php?id=53021.

Choi, Ki Sueng, Patricio Riva-Posse, Robert E. Gross, and Helen S. Mayberg. "Mapping the 'Depression Switch' During Intraoperative Testing of Subcallosal Cingulate Deep Brain Stimulation." *JAMA Neurology* 72, no. 11 (November 2015): 1252–60. doi:10.1001/jamaneurol.2015.2564.

Dobbs, David. "A Depression Switch?" *New York Times Magazine*. April 2, 2006.

Faria, Miguel A. "Violence, Mental Illness, and the Brain: A Brief History of Psychosurgery: Part 3—From Deep Brain Stimulation to Amygdalotomy for Violent Behavior, Seizures, and Pathological Aggression in Humans." *Surgical Neurology International* 4 (2013): 91. doi:10.4103/2152-7806.115162.

Franzini, Angelo, Giovanni Broggi, Roberto Cordella, Ivano Dones, and Giuseppe Messina. "Deep-Brain Stimulation for Aggressive and Disruptive Behavior." *World Neurosurgery* 80, no. 3–4 (2013): S29.e11-4. doi:10.1016/j.wneu.2012.06.038.

Fumagalli, Manuela, and Alberto Priori. "Functional and Clinical Neuroanatomy of Morality." *Brain : A Journal of Neurology* 135, no. Pt 7 (July 2012): 2006–21. doi:10.1093/brain/awr334.

"Gain on Schizophrenia?: Two Volunteers." *New York Times*, May 6, 1956.

Gilbert, Frédéric, and Daniela Ovadia. "Deep Brain Stimulation in the Media: Over-Optimistic Portrayals Call for a New Strategy Involving Journalists and Scientists in Ethical Debates." *Frontiers in Integrative Neuroscience* 5 (2011): 16. doi:10.3389/fnint.2011.00016.

Gkotsi, Georgia-Martha, and Lazare Benaroyo. "Neuroscience and the Treatment of Mentally Ill Criminal Offenders: Some Ethical Issues." *Journal of Ethics in Mental Health* 6 (2014) no. Supplement: Neuroethics (n.d.).

Hariz, Marwan, Patric Blomstedt, and Ludvic Zrinzo. "Deep Brain Stimulation between 1947 and 1987: The Untold Story." *Neurosurgical Focus* 29, no. 2 (August 2010): E1. doi:10.3171/2010.4.FOCUS10106.

———. "Future of Brain Stimulation: New Targets, New Indications, New Technology." *Movement Disorders : Official Journal of the Movement Disorder Society* 28, no. 13 (November 2013): 1784–92. doi:10.1002/mds.25665.

Harrison, Emma. "Mental Disorder Is Induced in Test: Scientists Report Developing Schizophrenia Symptoms in 2 'Normal' Persons Hypothesis Explained." *New York Times*, May 4, 1956.

Heath, Robert G. "Correlation of Brain Activity with Emotion: A Basis for Developing Treatment of Violent-Aggressive Behavior." *The Journal of the American Academy of Psychoanalysis* 20, no. 3 (1992): 335–46.

Heath, Robert G. "Correlation of Brain Function with Emotional Behavior." *Biological Psychiatry* 11, no. 4 (August 1976): 463–80.

———. "Fastigial Nucleus Connections to the Septal Region in Monkey and Cat: A Demonstration with Evoked Potentials of a Bilateral Pathway." *Biological Psychiatry* 6, no. 2 (April 1973): 193–96.

———. "The Human Brain: Instrument of Progress or Disaster?" *Biological Psychiatry.* United States, September 1985.

———. "Modulation of Emotion with a Brain Pacemaker: Treatment for Intractable Psychiatric Illness." *The Journal of Nervous and Mental Disease* 165, no. 5 (November 1977): 300–317.

———. "Pleasure and Brain Activity in Man: Deep and Surface Electroencephalograms During Orgasm." *The Journal of Nervous and Mental Disease* 154, no. 1 (January 1972): 3–18.

———. "Statement of Robert G. Heath." In *Quality of Health Care—Human Experimentation, 1973 : Hearings Before the Subcommittee on Health of the Committee on Labor and Public Welfare, United States Senate, Ninety-Third Congress, First Session, on S. 974 . . . ,* edited by United States Congress, Senate Committee on Labor and Public Welfare. Subcommittee on Health. Washington, DC, 1973.

———. *Studies in Schizophrenia: A Multidisciplinary Approach to Mind-Brain Relationship*, 1954.

Heath, Robert G., and Floris De Balbian Verster. "Effects of Chemical Stimulation to Discrete Brain Areas." *The American Journal of Psychiatry* 117 (May 1961): 980–90. doi:10.1176/ajp.117.11.980.

Heath, Robert G., and Iris M. Krupp. "Catatonia Induced in Monkeys by Antibrain Antibody." *The American Journal of Psychiatry* 123, no. 12 (June 1967): 1499–1504. doi:10.1176/ajp.123.12.1499.

———. "Schizophrenia as an Immunologic Disorder: I. Demonstration of Antibrain Globulins by Fluorescent Antibody Techniques." *Archives of General Psychiatry* 16, no. 1 (January 1, 1967): 1–9.

Heath, Robert G., Alvin M. Rouchell, Raeburn C. Llewellyn, and Cedric F. Walker. "Cerebellar Pacemaker Patients: An Update." *Biological Psychiatry* 16, no. 10 (October 1981): 953–62.

Heath, Robert G., Aris W. Cox, and Leonard S. Lustick. "Brain Activity During Emotional States." *The American Journal of Psychiatry* 131, no. 8 (August 1974): 858–62. doi:10.1176/ajp.131.8.858.

Heath, Robert G., and Cedric F. Walker. "Correlation of Deep and Surface Electroencephalograms with Psychosis and Hallucinations in Schizophrenics: A Report of Two Cases." *Biological Psychiatry* 20, no. 6 (June 1985): 669–74.

Heath, Robert G., Colby W. Dempesy, C. J. Fontana, and A. T. Fitzjarrell. "Feedback Loop Between Cerebellum and Septal-Hippocampal Sites: Its Role in Emotion and Epilepsy." *Biological Psychiatry* 15, no. 4 (August 1980): 541–56.

Heath, Robert G., Denis E. Franklin, and David Shraberg. "Gross Pathology of the Cerebellum in Patients Diagnosed and Treated as Functional Psychiatric Disorders." *The Journal of Nervous and Mental Disease* 167, no. 10 (October 1979): 585–92.

Heath, Robert G., Denis E. Franklin, Cedric F. Walker, and James W. Keating Jr. "Cerebellar Vermal Atrophy in Psychiatric Patients." *Biological Psychiatry* 17, no. 5 (May 1982): 569–83.

Heath, Robert G., Iris M. Krupp, Lawrence W. Byers, and Jan I. Liljekvist. "Schizophrenia as an Immunologic Disorder: II. Effects of Serum Protein Fractions on Brain Function." *Archives of General Psychiatry* 16, no. 1 (January 1, 1967): 10–23.

———. "Schizophrenia as an Immunologic Disorder. III. Effects of Antimonkey and Antihuman Brain Antibody on Brain Function." *Archives of General Psychiatry* 16, no. 1 (January 1967): 24–33.

Heath, Robert G., Raeburn C. Llewellyn, and Alvin M. Rouchell. "The Cerebellar Pacemaker for Intractable Behavioral Disorders and Epilepsy: Follow-up Report." *Biological Psychiatry* 15, no. 2 (April 1980): 243–56.

Hooper, Judith. "Brain Pacemakers: Pleasure on Command: Robert G. Heath." *Omni* (April 1984).

Horgan, John. "The Myth of Mind Control: Will Anyone Ever Decode the Human Brain?" *Discover* 25, no. 10 (2004): 40–47.

———. "What Are Science's Ugliest Experiments?" *Scientific American Blog*, May 14, 2012, https://blogs.scientificamerican.com/cross-check/what-are-sciences-ugliest-experiments/.

Horrock, Nicholas M. "Private Institutions Used in C.I.A. Effort to Control Behavior." *New York Times*, August 2, 1977.

Houser, H. "Treatment for the Acute Mentally Ill." *The Tulanean* (July 1959).

Jones, Amanda L., Bryan J. Mowry, Michael P. Pender, and Judith M. Greer. "Immune Dysregulation and Self-Reactivity in Schizophrenia: Do Some Cases of Schizophrenia Have an Autoimmune Basis?" *Immunology and Cell Biology* 83, no. 1 (February 2005): 9–17. doi:10.1111/j.1440-1711.2005.01305.x.

Jones, Dan. "Morality 2.0 : How Manipulating Our Minds Could Save the World." *New Scientist* 227, no. 3040 (2015): 36(4).

Kety, Seymour S. "Biochemical Theories of Schizophrenia." *Science* 129, no. 3363 (June 12, 1959): 1590 LP-1596.

Khandaker, Golam M., Lesley Cousins, Julia Deakin, Belinda R. Lennox, Robert Yolken, and Peter B. Jones. "Inflammation and Immunity in Schizophrenia: Implications for Pathophysiology and Treatment." *The Lancet Psychiatry* 2, no. 3 (March 2015): 258–70. doi:10.1016/S2215-0366(14)00122-9.

Kline, Nathan S., and Eugene Laska. *Computers and Electronic Devices in Psychiatry*. New York [usw.]: Grune & Stratton, 1968.

Konarski, Jakub Z., Roger S. McIntyre, Larry A. Grupp, and Sidney H. Kennedy. "Is the Cerebellum Relevant in the Circuitry of Neuropsychiatric Disorders?" *Journal of Psychiatry & Neuroscience : JPN* 30, no. 3 (May 2005): 178–86.

Kroken, Rune A., Else-Marie Loberg, Tore Dronen, Renate Gruner, Kenneth Hugdahl, Kristiina Kompus, Silje Skrede, and Erik Johnsen. "A Critical Review of Pro-Cognitive Drug Targets in Psychosis: Convergence on Myelination and Inflammation." *Frontiers in Psychiatry* 5 (2014): 11. doi:10.3389/fpsyt.2014.00011.

Kuhn, Jens, Christian P. Bührle, Doris Lenartz, and Volker Sturm. "Deep Brain Stimulation in Addiction Due to Psychoactive Substance Use." *Handbook of Clinical Neurology* 116 (2013): 259–69. doi:10.1016/B978-0-444-53497-2.00021-8.

Kuhn, Jens, M. Möller, J. F. Treppmann, Christian Bartsch, Doris Lenartz, Theo O. J. Gruendler, M. Maarouf, et al. "Deep Brain Stimulation of the Nucleus Accumbens and Its Usefulness in Severe Opioid Addiction." *Molecular Psychiatry* (February 2014). doi:10.1038/mp.2012.196.

Laxton, Adrian W., David F. Tang-Wai, Mary Pat McAndrews, Dominik Zumsteg, Richard Wennberg, Ron Keren, John Wherrett, et al. "A Phase I Trial of Deep Brain Stimulation of Memory Circuits in Alzheimer's Disease." *Annals of Neurology* 68, no. 4 (October 2010): 521–34. doi:10.1002/ana.22089.

Liao, S. Matthew. "Could Deep Brain Stimulation Fortify Soldiers' Minds?" *Scientific American Blog*, September 4, 2014. https://blogs.scientificamerican.com/mind-guest-blog/could-deep-brain-stimulation-fortify-soldiers-minds/.

Lipsman, Nir, and Andres M. Lozano. "Cosmetic Neurosurgery, Ethics, and Enhancement." *The Lancet Psychiatry* (July 2015). doi:10.1016/S2215-0366(15)00206-0.

Lozano, Andres M., and Helen S. Mayberg. "Treating Depression at the Source." *Scientific American* 312, no. 2 (2015): 68. doi:10.1038/scientificamerican0215-68.

Maley, Jason H., Jorge E. Alvernia, Edison P. Valle, and Donald Richardson. "Deep Brain Stimulation of the Orbitofrontal Projections for the Treatment of Intermittent Explosive Disorder." *Neurosurgical Focus* 29, no. 2 (August 2010): E11. doi:10.3171/2010.5.FOCUS10102.

Mantione, Mariska, Martijn Figee, and Damiaan Denys. "A Case of Musical Preference for Johnny Cash Following Deep Brain Stimulation of the Nucleus Accumbens." *Frontiers in Behavioral Neuroscience* 8 (2014): 152. doi:10.3389/fnbeh.2014.00152.

Mark, Vernon H., and Frank R. Ervin. *Violence and the Brain*. New York [etc.]: Harper and Row, 1970.

Mikell, Charles B., Guy M. McKhann, Solomon Segal, Robert A. McGovern, Matthew B. Wallenstein, and Holly Moore. "The Hippocampus and Nucleus Accumbens as Potential Therapeutic Targets for Neurosurgical Intervention in Schizophrenia." *Stereotactic and Functional Neurosurgery* 87, no. 4 (2009): 256–65. doi:10.1159/000225979.

"Mind and Antibody: The Return of Immunopsychiatry." *The Lancet Psychiatry* (March 2015). doi:10.1016/S2215-0366(15)00057-7.

Moan, Charles E., and Robert G. Heath. "Septal Stimulation for the Initiation of Heterosexual Behavior in a Homosexual Male." *Journal of Behavior Therapy and Experimental Psychiatry* 3, no. 1 (1972): 23–30. doi:10.1016/0005-7916(72)90029-8.

Montgomery, Erwin B. Jr. "The Epistemology of Deep Brain Stimulation and Neuronal Pathophysiology." *Frontiers in Integrative Neuroscience* 6 (2012): 78. doi:10.3389/fnint.2012.00078.

Nadjari, Douglas. "Hackney Releases CIA : Tulane Statement Reveals Drug Tests on Human Volunteer." *The Tulane Hullabaloo*, March 31, 1978.

Parvizi, Josef, Vinitha Rangarajan, William R. Shirer, Nikita Desai, and Michael D. Greicius. "The Will to Persevere Induced by Electrical Stimulation of the Human Cingulate Gyrus." *Neuron* 80, no. 6 (December 2013): 1359–67. doi:10.1016/j.neuron.2013.10.057.

Paul, Steven M., Robert G. Heath, and Jeffrey P. Ellison. "Histochemical Demonstration of a Direct Pathway from the Fastigial Nucleus to the Septal Region." *Experimental Neurology* 40, no. 3 (September 1973): 798–805.

Portenoy, Russell K., Jens O. Jarden, John J. Sidtis, Richard B. Lipton, Kathleen M. Foley, and David A. Rottenberg. "Compulsive Thalamic Self-Stimulation: A Case with Metabolic, Electrophysiologic and Behavioral Correlates." *Pain* 27, no. 3 (December 1986): 277–90.

Ruff, Christian C., Giuseppe Ugazio, and Ernst Fehr. "Changing Social Norm Compliance with Noninvasive Brain Stimulation." *Science* 342, no. 6157 (October 2013): 482–84. doi:10.1126/science.1241399.

Rushton, Bill. "The Mysterious Experiments of Dr. Heath: In Which We Wonder Who Is Crazy & Who Is Sane." *Courier*, September 4, 1974.

Schläpfer, Thomas E., Bettina H. Bewernick, Sarah Kayser, Burkhard Madler, and Volker A. Coenen. "Rapid Effects of Deep Brain Stimulation for Treatment-Resistant Major Depression." *Biological Psychiatry* 73, no. 12 (June 2013): 1204–12. doi:10.1016/j.biopsych.2013.01.034.

Schläpfer, Thomas E., Bettina H. Bewernick, Sarah Kayser, Rene Hurlemann, and Volker A. Coenen. "Deep Brain Stimulation of the Human Reward System for Major Depression-Rationale: Outcomes and Outlook." *Neuropsychopharmacology: Official Publication of the American College of Neuropsychopharmacology* 39, no. 6 (May 2014): 1303–14. doi:10.1038/npp.2014.28.

Smith, Gwenn S., Adrian W. Laxton, David F. Tang-Wai, Mary Pat McAndrews, Andreea O. Diaconescu, Clifford I. Workman, and Andres M. Lozano. "Increased Cerebral Metabolism After 1 Year of Deep Brain Stimulation in Alzheimer Disease." *Archives of Neurology* 69, no. 9 (September 2012): 1141–48. doi:10.1001/archneurol.2012.590.

Steiner, Johann, Martin Walter, Wenzel Glanz, Zoltan Sarnyai, Hans-Gert Bernstein, Stefan Vielhaber, Andrea Kastner, et al. "Increased Prevalence of Diverse N-Methyl-D-Aspartate Glutamate Receptor Antibodies in Patients with an Initial Diagnosis of Schizophrenia: Specific Relevance of IgG NR1a Antibodies for Distinction from N-Methyl-D-Aspartate Glutamate Receptor Encephalitis." *JAMA Psychiatry* 70, no. 3 (March 2013): 271–78. doi:10.1001/2013.jamapsychiatry.86.

Sturm, Volker, Oliver Fricke, Christian P. Bührle, Doris Lenartz, Mohammad Maarouf, Harald Treuer, Jurgen K. Mai, and Gerd Lehmkuhl. "DBS in the Basolateral Amygdala Improves Symptoms of Autism and Related Self-Injurious Behavior: A Case Report and Hypothesis on the Pathogenesis of the Disorder." *Frontiers in Human Neuroscience* 6 (2012): 341. doi:10.3389/fnhum.2012.00341.

Synofzik, Matthis, Thomas E. Schläpfer, and Joseph J. Fins. "How Happy Is Too Happy?: Euphoria, Neuroethics, and Deep Brain Stimulation of the Nucleus Accumbens." *AJOB Neuroscience* 3, no. 1 (January 1, 2012): 30–36. doi:10.1080/21507740.2011.635633.

Tomlinson, Wallace K., and Robert Galbraith Heath. "An Attempt at Historical Perspective." New Orleans, LA: n.d.

"Transcript of Notes Taken at Meeting of Special Psychiatric Committee." New Orleans, LA: n.d.

"Tulane Professor Urged to Work for CIA." *Times-Picayune*, 1977.

United States National Commission for the Protection of Human Subjects of Biomedical and Behavioral Research. *Psychosurgery: Report and Recommendations.* Edited by Harold Alexander Abramson. DHEW Publication ; No. (OS)77-0001. The Commission, 1977.

Valencia-Alfonso, Carlos-Eduardo, Judy Luigjes, Ruud Smolders, Michael X. Cohen, Nina Levar, Ali Mazaheri, Pepijn van den Munckhof, P. Richard Schuurman, Wim van den Brink, and Damiaan Denys. "Effective Deep Brain Stimulation in Heroin Addiction: A Case Report with Complementary Intracranial Electroencephalogram." *Biological Psychiatry* (April 2012). doi:10.1016/j.biopsych.2011.12.013.

Voytek, Bradley. "The Most Unethical Study I've Ever Seen." *Quora*, August 7, 2011, https://www.quora.com/profile/Bradley-Voytek/Posts/The-most-unethical-study-Ive-ever-seen.

Widge, Alik S., Kristen K. Ellard, Angelique C. Paulk, Ishita Basu, Ali Yousefi, Samuel Zorowitz, Anna Gilmour, et al. "Treating Refractory Mental Illness with Closed-Loop Brain Stimulation: Progress Towards a Patient-Specific Transdiagnostic Approach." *Experimental Neurology* 287, no. Pt 4 (January 2017): 461–72. doi:10.1016/j.expneurol.2016.07.021.

Wu, Hemmings, Hartwin Ghekiere, Dorien Beeckmans, Tim Tambuyzer, Kris van Kuyck, Jean-Marie Aerts, and Bart Nuttin. "Conceptualization and Validation of an Open-Source Closed-Loop Deep Brain Stimulation System in Rat." *Scientific Reports* 4 (April 2015): 9921. doi:10.1038/srep09921.

Wu, Hemmings, Pieter Jan Van Dyck-Lippens, Remco Santegoeds, Kris van Kuyck, Loes Gabriels, Guozhen Lin, Guihua Pan, et al. "Deep-Brain Stimulation for Anorexia Nervosa." *World Neurosurgery* 80, no. 3–4 (2013): S29.e1-10. doi:10.1016/j.wneu.2012.06.039.

ACKNOWLEDGMENTS

This book would not have existed without the generous help and encouragement from a group of scientists and gentlemen who have all worked with and known Robert G. Heath. They often welcomed me into their homes and spent countless hours answering questions and delving into the past. I am forever indebted to the late Frank Ervin and Charles Fontana, and to Donald Richardson, Alan Lipton, James Eaton, Charles O'Brien, Joseph DiGiacomo, Colby "Skip" Dempesy, Don Gallant, Robert Begtrup, and Sam Bailine. And had it not been for Robert Heath and his wonderful family, I would not have had the chance to access the treasure trove consisting of films recording the world's first experiments with deep brain stimulation in patients.

I owe a very special thanks to David Merrick, who spoke candidly about being a patient receiving an unusual treatment, and to his sister Barbara Chester for being open about her family's most difficult struggle.

Thanks to Todd Ochs, Max Fink, and Elliott Valenstein for sharing their personal experiences and views. And to Alan Baumeister for providing me with his academic perspective and important source material.

I am deeply grateful to have been received by some of today's international pioneers in deep brain stimulation. Helen Mayberg, Thomas Schläpfer, Volker Coenen, Darin Dougherty, and Alik Widge all gracefully opened the doors to their labs and operating rooms.

I feel lucky to have had the gifted Robin Dennis offering invaluable advice and editing along the way. And I am forever thankful that my agent, Peter Tallack, has kept insisting that being Danish is no hindrance in the international book world—even when it sometimes is.

Profound thanks to Klaus Rothstein and Janne Breinholt Bak for their reading and critique of early versions of the manuscript and for their general encouragement.

As always, Russell Dees has been indispensable, his unfailing ear and gift for translation never cease to impress me. Librarian Dorte Nielsen provided generous help with organizing references.

To my editor, Stephen Morrow, whose support, guidance, and enthusiasm for the story has been wonderful—my deep-felt gratitude.

INDEX